1993

ADVANCED MATHEMATICAL THINKING

Mathematics Education Library

VOLUME 11

The titles published in this series are listed at the end of this volume.

ADVANCED MATHEMATICAL THINKING

Edited by

DAVID TALL
Science Education Department,
University of Warwick

KLUWER ACADEMIC PUBLISHERS
DORDRECHT / BOSTON / LONDON

Library of Congress Cataloging-in-Publication Data

Advanced mathematical thinking / edited by David Tall.
        p.   cm. -- (Mathematics education library ; v. 11)
    A project of the Working Group of PME.
    Includes bibliographical references (p.   ) and index.
    ISBN 0-7923-1456-5 (acid-free paper)
    1. Mathematics--Study and teaching. 2. Mathematics--Psychological
aspects.   I. Tall, David Orme.   II. Series.
    QA11.A36   1991
    510'.71--dc20   91-31945

ISBN 0-7923-1456-5

Published by Kluwer Academic Publishers,
P.O. Box 17, 3300 AA Dordrecht, The Netherlands.

Kluwer Academic Publishers incorporates
the publishing programmes of
D. Reidel, Martinus Nijhoff, Dr W. Junk and MTP Press.

Sold and distributed in the U.S.A. and Canada
by Kluwer Academic Publishers,
101 Philip Drive, Norwell, MA 02061, U.S.A.

In all other countries, sold and distributed
by Kluwer Academic Publishers Group,
P.O. Box 322, 3300 AH Dordrecht, The Netherlands.

Printed on acid-free paper

Printed in the Netherlands

TABLE OF CONTENTS

TABLE OF CONTENTS

I : THE NATURE OF
ADVANCED MATHEMATICAL THINKING

II: COGNITIVE THEORY
OF ADVANCED MATHEMATICAL THINKING

III : RESEARCH INTO THE TEACHING AND LEARNING OF ADVANCED MATHEMATICAL THINKING

EPILOGUE

PREFACE

Advanced Mathematical Thinking has played a central role in the development of human civilization for over two millennia. Yet in all that time the serious study of the nature of advanced mathematical thinking – what it is, how it functions in the minds of expert mathematicians, how it can be encouraged and improved in the developing minds of students – has been limited to the reflections of a few significant individuals scattered throughout the history of mathematics. In the twentieth century the theory of mathematical education during the compulsory years of schooling to age 16 has developed its own body of empirical research, theory and practice. But the extensions of such theories to more advanced levels have only occurred in the last few years.

In 1976 The International Group for the Psychology of Mathematics (known as PME) was formed and has met annually at different venues round the world to share research ideas. In 1985 a Working Group of PME was formed to focus on Advanced Mathematical Thinking with a major aim of producing this volume.

The text begins with an introductory chapter on the psychology of advanced mathematical thinking, with the remaining chapters grouped under three headings:

- the nature of advanced mathematical thinking,
- cognitive theory,

and

- reviews of the progress of cognitive research into different areas of advanced mathematics.

It is written in a style intended both for mathematicians and for mathematics educators, to encourage an interest in the cognitive difficulties experienced by students of the former and to extend the psychological theories of the latter through to later stages of development. We are cognizant of the fact that it is essential to understand the nature of the thinking of mathematical experts to see the full spectrum of mathematical growth. We therefore begin with an introductory chapter on the psychology of advanced mathematical thinking. This is followed by three chapters which focus on the nature of advanced mathematical thinking: a study of the mental processes involved, the essential qualities of mathematical creativity and the mathematician's view of proof.

The processes prove to be subtle and complex and, sadly, few of the more advanced processes are made available to the average student in an advanced mathematical course. Creativity is concerned with how the subtle ideas of research are built in the mind. Proof is how they are ordered in a logical development both to verify the nature of the relationships and also to present them for approval to the mathematical community.

However, there is a huge gulf between the way in which ideas are built cognitively and the way in which they are arranged and presented in a deductive order. This warns us that simply presenting a mathematical theory as a sequence of definitions, theorems and proofs (as happens in a typical university course) may show the logical structure of the mathematics, but it fails to allow for the psychological growth of the developing human mind.

We begin the part of the book on cognitive theory by considering the way in which formal mathematical definitions are conceived by students and how this can be at variance with the formal theory. As a result of mentally manipulating a (mathematical) concept an individual develops an idiosyncratic personal *concept image* which is the product of experience and mental activity. Empirical research shows how this can give rise to subtle conflicts that can cause *cognitive obstacles* in the mind of the developing student and act as a barrier to attaining the formal ideas in the theory. The next chapter looks at the mental objects that are the material of mathematical thought – the *conceptual entities* that are manipulated in the mind during advanced mathematical thinking, and how these entities are represented by different kinds of symbolism. The final chapter in this part considers how these conceptual entities are formed – through the process of *reflective abstraction*. All advanced mathematical concepts are "abstract". This chapter postulates a theory of how these concepts start as *processes* which are *encapsulated* as *mental objects* that are then available for higher level abstract thought. Such a theory can give insight into how mathematicians develop advanced mathematical ideas, yet may fail to pass these thinking processes on to students, and what might be done to improve the situation.

The remainder of the book is concerned with overviews of empirical research and theory in various specific topics. First the question of the nature of advanced mathematical thinking is addressed and how (if at all) it differs from more elementary thinking occurring in younger children. Then there follow chapters on functions, limits, analysis, infinity, proof, and the growing use of the computer in advanced mathematics. Each one of these reveals a wide variety of obstacles in students' mental imagery and often extremely limited conceptions of formal concepts which are the unforseen consequences of the manner in which the subject is presented to the student. A variety of more cognitively appropriate approaches are postulated, some with empirical evidence of success. These include:

- the participation of the student in the process of mathematical thinking through an active process of "scientific debate", rather than passive receipt of pre-organized theory,

- the direct confrontation of the student with conflict which occurs in developing new theoretical constructs, to help them reflect on the problem and build a new, more coherent, cognitive structure.

- the building up of appropriate intuitive foundations for the advanced mathematical concepts, through an approach which balances cognitive growth and an appreciation of logical development.

- the use of visualization, particularly utilizing a computer, to give the student an overall view of concepts and enabling more versatile methods of handling the information,

- the use of programming to cause the student to think through mathematical processes in a way which can be encapsulated by reflective abstraction.

In all these ways we believe that empirical research into advanced thinking processes related to complementary cognitive theory can have a significant effect in improving the education of students at an advanced level.

In every chapter the authors have been encouraged to impress their own personalities on their view of the phenomena, but this has been done within a framework of internal consultation. Each participant operates from personal constructs within a context of mutual support and constructive criticism from other authors and the final manuscript has been recast by the editor to enable it to be read throughout as a single text rather than as a collection of disconnected papers. This was made possible through the wonders of modern technology, using a Macintosh SE/30 computer to enable the editor to redraft the chapters and set the whole book as camera-ready copy.

The cognitive theory of advanced mathematical thinking is developing apace. This study is the first step in making the broad sweep of current ideas in the advanced mathematical education community available to a wider readership.

David Tall

ACKNOWLEDGEMENTS

In preparing this book we acknowledge the assistance and support of a wide range of mathematicians and mathematics educators. First we thank the International Committee of PME for encouraging us to meet each year as a continuing Working Group of PME. Then we thank those members of the Working Group who do not appear as authors but who have helped in many ways – presenting papers, organizing and reporting sessions – including Janet Duffin, Ulla Kürstein-Jensen, Miriam Amit, Anna Sfard, Janine Rogalski. Within the group of authors, all have participated in criticizing and encouraging the writing of others. However, it is with grateful thanks that the editor wishes to acknowledge assistance beyond the call of duty from Tommy Dreyfus, Ted Eisenberg, Ed Dubinsky, and particularly from Gontran Ervynck, whose enthusiasm generated the original impetus for the creation of this book.

INTRODUCTION

CHAPTER 1

THE PSYCHOLOGY
OF
ADVANCED MATHEMATICAL THINKING

DAVID TALL

In the opening chapter of *The Psychology of Invention in the Mathematical Field*, the mathematician Jacques Hadamard highlighted the fundamental difficulty in discussing the nature of the psychology of advanced mathematical thinking:

> ... that the subject involves two disciplines, psychology and mathematics, and would require, in order to be treated adequately, that one be both a psychologist and a mathematician. Owing to the lack of this composite equipment, the subject has been investigated by mathematicians on the one side, by psychologists on the other ...   (Hadamard, 1945, page 1.)

Exponents of the two disciplines are likely to view the subject in different ways – the psychologist to extend psychological theories to thinking processes in a more complex knowledge domain – the mathematician to seek insight into the creative thinking process, perhaps with the hope of improving the quality of teaching or research. Although we will consider the nature of advanced mathematical thinking from a psychological viewpoint, our main aim will be to seek insights of value to the mathematician in his professional work as researcher and teacher.

We begin by looking at pertinent psychological considerations which will lay the foundations for ideas introduced not only in the remainder of the chapter, but in the book as a whole. We then focus our attention on the full cycle of activity in advanced mathematical thinking: from the creative act of considering a problem context in mathematical research that leads to the creative formulation of conjectures and on to the final stage of refinement and proof. We postulate that many of the activities that occur in this cycle also occur in elementary mathematical problem-solving, but the possibility of formal definition and deduction is one factor which distinguishes advanced mathematical thinking. We will also find that teaching undergraduate mathematics often presents the final form of the deduced theory rather than enabling the student to participate in the full creative cycle. In the words of Skemp (1971), current approaches to undergraduate teaching tend to give students the *product of mathematical thought* rather than the *process of mathematical thinking*.

Not only may current methods of presenting advanced mathematical knowledge fail to give the full power of mathematical thinking, it also has another, equally serious, deficiency: *a logical presentation may not be appropriate for the cognitive development of the learner*. Indeed, much of the empirical theory reported in the later chapters of the book reveals cognitive obstacles which arise as students struggle to come to terms with ideas which challenge and contradict their current knowledge structure. Fortunately, we are also able to report empirical evidence that appropriate sequences of learning and instruction designed to help the student actively construct the concepts can prove highly successful.

1. COGNITIVE CONSIDERATIONS

We begin by looking, not at the logic and order of the public evidence of mathematical thinking found in research articles and text-books, but at the way in which these coherent relationships are built in mathematical research and implications for how this might be implemented in teaching and learning.

1.1 DIFFERENT KINDS OF MATHEMATICAL MIND

Writing in the first decade of this century, the celebrated mathematician Henri Poincaré asserted:

It is impossible to study the works of the great mathematicians, or even those of the lesser, without noticing and distinguishing two opposite tendencies, or rather two entirely different kinds of minds. The one sort are above all preoccupied with logic; to read their works, one is tempted to believe they have advanced only step by step, after the manner of a Vauban[1] who pushes on his trenches against the place besieged, leaving nothing to chance. The other sort are guided by intuition and at the first stroke make quick but sometimes precarious conquests, like bold cavalrymen of the advanced guard.   (Poincaré, 1913, p. 210)

He supported his arguments by contrasting the work of various mathematicians, including the famous German analysts, Weierstrass and Riemann, relating this to the work of students:

Weierstrass leads everything back to the consideration of series and their analytic transformations; to express it better, he reduces analysis to a sort of prolongation of arithmetic; you may turn through all his books without finding a figure. Riemann, on the contrary, at once calls geometry to his aid; each of his conceptions is an image that no one can forget, once he has caught its meaning.

... Among our students we notice the same differences; some prefer to treat their problems 'by analysis', others 'by geometry'. The first are incapable of 'seeing in space', the others are quickly tired of long calculations and become perplexed.   (Poincaré, 1913, p. 212)

Of course, there are not just two different kinds of mathematical mind, but many. Kronecker agreed with Weierstrass that logical proof was of paramount importance and transcended intuitive visual arguments, but their fundamental beliefs in the nature of mathematical concepts were very different. Weierstrass declared that "an irrational number has as real an existence as anything else in the world of concepts", but Kronecker was unable to accept the actual infinity of real numbers, asserting that "God gave us the integers, the rest is the work of man". Based on the Weierstrassian notion of the actual infinity of real numbers, Cantor was able to produce an infinite counting argument to show that there are strictly "more" real numbers than algebraic numbers (solutions of polynomial equations with integer coefficients). He therefore claimed that there exists a real non-algebraic number, without giving an explicit method to construct one. This was anathema to Kronecker who caused Cantor's paper to be rejected from publication in Crelle's Journal in 1873.

---

[1] Sebastien de Vauban (1633-1707) was a French military engineer who revolutionized the art of siege craft and defensive fortifications.

Such arguments about the foundations of mathematics led to the development of several different strands of mathematical philosophy at the beginning of the twentieth century. The *intuitionist* view represented by Kronecker asserted that mathematical concepts only exist when their construction is demonstrated from the integers, the *formalist* view of Hilbert affirmed that mathematics is the meaningful manipulation of meaningless marks written on paper, whilst the *logicist* view of Russell, declared that mathematics consists of deductions using the laws of logic.

Practising mathematicians tend to distance themselves from esoteric arguments and simply get on with their work of stating and proving theorems. Thus the twentieth century has seen the demise of Kronecker's views and the triumph of a pragmatic mixture of formalism and logic. It has seen the creation of a large number of formal systems based on logical deduction from formal definitions and axioms – an approach that survived the apparently mortal blow struck by Gödel's incompleteness theorem, that any axiomatic system including the integers must contain true statements that cannot be proved by a finite sequence of steps within the system.

The textbook by Bishop (1967) on constructive analysis – which insists on algorithmic construction proofs and disallows proof by contradiction alone – seems but an isolated singularity in the dynamic flow of twentieth century mathematical creativity.

Nevertheless, the recent introduction of computer technology may yet see a new renaissance in constructibility because of the way that computers manipulate data:

Computers have affected mathematics as inevitably as the development of railroads affected patterns of land development. With computers it is possible to test hypotheses and compile data with ease that formerly would have been accessible, if at all, only via the most sophisticated techniques. This has affected not only the sort of questions that mathematicians work on, but the very way that they think. One has to ask oneself which examples can be tested on a computer, a question which forces one to consider concrete algorithms and to try to make them efficient. Because of this and because algorithms have real-life applications of considerable importance, the development of algorithms has become a respectable topic in its own right.   (Edwards, 1987)

The reason for raising these differences in mathematician's perceptions is to heighten the readers' awareness of their own part in life's rich tapestry, with a personal view of mathematics that will differ in many ways from the conceptions of others. It may come as a surprise when one first realizes that other people have radically different thinking processes. It happened to the author when using pictures to help students visualize ideas in mathematical analysis, at a time when he did not question the implicit belief that such an approach was universally valid. Whilst writing a textbook on complex analysis, a colleague in the next room was engaged on a similar enterprise, yet the latter's book had almost no pictures at all. He only included a diagram illustrating the argument of a complex number after a great deal of heart searching. To him a real number was an element of a complete ordered field (satisfying specific axioms) and a complex number was an ordered pair of real numbers. The argument of a complex number (x,y) was defined as a real number α such that

$$\cos(\alpha) = \frac{x}{\sqrt{x^2+y^2}}, \quad \sin(\alpha) = \frac{y}{\sqrt{x^2+y^2}}$$

where sin and cos were defined by real power series. The theory did not require a geometrical meaning. He took this hard line to make sure that his arguments were the product of logical deduction and not dependent anywhere on geometric intuition. At the time the author was sympathetic to this philosophical viewpoint, but considered it too sophisticated for students. It was some considerable time later that the realization dawned that not all students shared the geometric point of view. No one view holds universal sway.

1.2 META-THEORETICAL CONSIDERATIONS

The discussion of the preceding session is a salutary reminder that any theory of the psychology of learning mathematics must take into account not only the growing conceptions of the students, but the conceptions of mature mathematicians. Mathematics is a shared culture and there are aspects which are context dependent. For example, an analyst's view of a differential may be very different from that of an applied mathematician, and a given individual may strike up different attitudes to this concept depending on whether it is considered in an analytic or applied context. We will see (chapter 11) that such attitudes can cause conflicts in students too.

At a far deeper psychological level we all have subtly different ways of viewing a given mathematical concept, depending on our previous experiences. For example, the "completeness axiom" for the real numbers is viewed by some as "filling in all the gaps between the rational numbers to give all the points on the number line". Such a view may imply that there is "no room" to fit in any more numbers: the number line is now "complete". The "real" number line, in particular cannot contain "infinitesimals" which are smaller than any positive rational yet not zero. But, for others, "completion" is only a technical axiom to adjoin the limit points of cauchy sequences of rational numbers. In this case it is perfectly possible to embed the real numbers in a variety of larger number fields, which include infinitesimals and infinite numbers. It is this view which leads to the modern infinitesimal theory of "non-standard analysis". The latter idea, however, is anathema to many mathematicians, including Cantor, who denied the existence of infinitesimals on the grounds that it was not possible to calculate the reciprocal of an infinite number in his theory of cardinal infinities. Even today many mathematicians are troubled by the infinitesimal ideas of non-standard analysis; they may not deny its logic, but they sense a deep-seated psychological unease as to its validity.

Thus any theory of the psychology of mathematical thinking must be seen in the wider context of human mental and cultural activity. There is not one true, absolute way of thinking about mathematics, but diverse culturally developed ways of thinking in which various aspects are relative to the context.

1.3 CONCEPT IMAGE AND CONCEPT DEFINITION

In Tall & Vinner (1981), the distinction is made between the individual's way of thinking of a concept and its formal definition, thus distinguishing between mathematics as a mental activity and mathematics as a formal system. This theory applies to expert mathematicians as well as developing students:

The human brain is not a purely logical entity. The complex manner in which it functions is often at variance with the logic of mathematics. It is not always pure logic that gives us insight, nor is it chance that makes us make mistakes... We shall use the term *concept image* to describe the total cognitive structure that is associated with the concept, which includes all the mental pictures and associated properties and processes. It is built up over the years through experiences of all kinds, changing as the individual meets new stimuli and matures. ... As the concept image develops it need not be coherent at all times. The brain does not work that way. Sensory input excites certain neuronal pathways and inhibits others. In this way different stimuli can activate different parts of the concept image, developing them in a way which need not make a coherent whole. (Tall & Vinner 1981)

In this way it is possible for conflicting views to be held in the mind of a given individual and to be evoked at different times without the individual being aware of the conflict until they are evoked simultaneously.

The mature mathematician is not immune from internal conflicts, but he or she has been able to link together large portions of knowledge into sequences of deductive argument. To such a person it seems so much easier to categorize this knowledge in a logically structured way. Thus a mature mathematician may consider it helpful to present material to students in a way which highlights the logic of the subject. However, a student without the experience of the teacher may find a formal approach initially difficult, a phenomenon which may be viewed by the teacher as a lack of experience or intellect on the part of the student. This is a comforting viewpoint to take, especially when the teacher is part of a mathematical community who share the mathematical understanding. But it is not realistic in the wider context of the needs of the students. What is essential – for them – is an approach to mathematical knowledge that grows as they grow: a cognitive approach that takes account of the development of their knowledge structure and thinking processes. To become mature mathematicians at an advanced level, they must ultimately gain insight into the ways of advanced mathematicians but, en route, they may find a stony path that will require a fundamental transition in their thinking processes.

1.4 COGNITIVE DEVELOPMENT

There are many competing theories in psychology. Behaviourist theory, built on external observation of stimulus and response, refuses to speculate about the internal workings of the mind. It provides observable and repeatable evidence of the behaviour of animals, including humans, under repeated stimuli, but it has limited application to mathematical thinking beyond the mechanics of routine algorithms. Constructivist psychology, on the other hand, attempts to discuss how mental ideas are created in the mind of each individual. This may pose a dialectic problem for the mathematician with a Platonic ideal of mathematics existing independently of the human mind, but it proves to give significant insight into the creative processes of research mathematicians as well as the difficulties experienced by mathematics students.

The great Swiss psychologist Piaget saw the individual's need to be in dynamic equilibrium with his environment as an underlying theme in his work. This equilibrium could be disturbed through the confrontation with new knowledge that conflicted with the old, and so a transition period might occur in which the knowledge structure is reconstructed to give a more mature level of equilibrium.

Piaget saw the child grow into the adult through a series of stages of equilibrium, each one richer than the one before. He identified four main stages. The first is the *sensori-motor* stage prior to the development of meaningful speech, followed by a *pre-operational* stage when the young child realizes the permanence of objects, which continue to exist even if they are temporarily out of sight. The child then goes through a transition into the period of *concrete operations* where he or she can stably consider concepts which are linked to physical objects, thence passing into a period of *formal operations* in the early teens when the kind of hypothetical "if–then" becomes possible.

Piagetian stage theory has been extended to higher levels to encompass advanced mathematical thinking. For instance, Ellerton (1985) suggested that Piaget's cycle of sensori-motor, pre-operational and concrete is the first level of a spiral cognitive development in which the formal stage is the beginning of another cycle of the same type at a higher level of abstraction. Biggs & Collis (1982) suggested a repetition of formal operations at successively higher levels, each repeating the learning cycle: unistructural, multistructural, relational.

A difficulty of applying such theory to college mathematics teaching is that many – probably most – college students are not able to perform at the abstract level of formal operations, which Piaget reported occurring in children during their early teens. Ausubel criticized the stage theory:

... because such a high percentage of American high school and college students fail to reach this abstract level of cognitive logical operations.   (Ausubel *et al* 1968, p. 230)
    Representative studies have indicated that only 15% of junior high school students ... 13.2% of high school students ... and 22% of college students were at this level.   (*ibid*, p. 238)

The concrete/formal distinction has proved to be a useful starting point in developing local hierarchies of difficulty in extensive studies such as Hart (1981) in the 11 to 16 age range, and the development of early calculus concepts by Orton (1980). But a significant failure of Piaget's stage theory for the design of new teaching strategies is his own assertion that the movement from one stage to another cannot be greatly accelerated by the affects of teaching. Differences of cognitive demand have often been used in a *negative* sense to describe students' difficulties, but rarely to provide *positive* criteria for designing new approaches to the subject. Papert (1980) asserted:

The Piaget of stage theory is essentially conservative, almost reactionary, in emphasizing what children cannot do. I strive to uncover a more revolutionary Piaget, one whose epistemological ideas might expand the known bounds of the human mind.

Advanced mathematics provides us with a useful metaphor which expands the vision of stage theory to a theory more valuable in the development of advanced mathematical thinking. Piaget used an analogy with group theory to underpin his sense of the dynamic equilibrium of cognitive growth. He saw the identity element as representing the stable state, and noted that stability could be maintained if any transformation from this state could be reversed, thus suggesting a group structure in which every element has an inverse. But the maintenance of a dynamic state of equilibrium has a more obvious mathematical metaphor in dynamical systems and catastrophe theory. Here a system controlled by continuously varying parameters can suddenly leap from one position of equilibrium to another when the first becomes untenable. Depending on the history of the varying parameters, the transition may be smooth, or it may be discontinuous. This analogy suggests that stage theory may just be a linear trivialization of a far more complex system of change, at least this may be so when the possible routes through a network of ideas become more numerous, as happens in advanced mathematical thinking.

1.5 TRANSITION AND MENTAL RECONSTRUCTION

A far more valuable aspect of Piaget's theory is the process of *transition* from one mental state to another. During such a transition, unstable behaviour is possible, with the experience of previous ideas conflicting with new elements. Piaget uses the terms *assimilation* to describe the process by which the individual takes in new data and *accommodation* the process by which the individual's cognitive structure must be modified. He sees assimilation and accommodation as complementary. During a transition much accommodation is required. Skemp (1979) puts similar ideas in a different way by distinguishing between the case where the learning process causes a simple *expansion* of the individual's cognitive structure and the case where there is cognitive conflict, requiring a mental *reconstruction*. It is this process of reconstruction which provokes the difficulties that occur during a transition phase.

Such transitions occur often in advanced mathematics as the individual struggles with new knowledge structure. Conflict is a phenomenon well-known to the mathematical mind.

1.6 OBSTACLES

The most serious problem occurs when the new ideas are not satisfactorily accommodated. In this case it may be possible for conflicting ideas to be present in an individual at one and the same time:

New knowledge often contradicts the old, and effective learning requires strategies to deal with such conflict. Sometimes the conflicting pieces of knowledge can be reconciled, sometimes one or the other must be abandoned, and sometimes the two can both be "kept around" if safely maintained in separate compartments.

(Papert, 1980, p. 121)

The thesis of Cornu (1983) studies the conceptual development of the limit process from school to university and underlines how the colloquial use of the term "limit" effects the mathematical usage. He discusses the notion of an "obstacle", introduced by Gaston Bachelard (1938):

An obstacle is a piece of knowledge; it is part of the knowledge of the student. This knowledge was at one time generally satisfactory in solving certain problems. It is precisely this satisfactory aspect which has anchored the concept in the mind and made it an obstacle. The knowledge later proves to be inadequate when faced with new problems and this inadequacy may not be obvious.

(Cornu 1983, (original in French))

The obstacles found by Cornu include the problems student face when they must calculate limits using techniques more subtle than simple numerical and algebraic operations. He discusses how the concept of infinity is introduced and is "surrounded in mystery", yet the new techniques "work" without the students understanding why. He demonstrates how students' experiences can lead to belief in the infinitely large and the infinitely small, with "nought point nine recurring" being a number "just less than one" and the symbol ε representing to many students a quantity that is smaller than any positive real number, but not zero. There are implicit assumptions that the limiting process "goes on forever", that the limit "can never be attained". (See chapter 10.)

Tall (1986a) suggests an explanation is given for these phenomena as the *generic extension principle*:

If an individual works in a restricted context in which all the examples considered have a certain property, then, in the absence of counter-examples, the mind assumes the known properties to be implicit in other contexts.

For example, most convergent sequences described to beginning students are of a simple kind given by a formula such as $1/n$, which tends to the limit (in this case zero), but the terms never *equal* the limit. In the absence of any counter-examples students begin to believe that this is always so. The rich experience of colloquial language supports this belief (Schwarzenberger & Tall, 1978), with phrases like "gets close to" suggesting that the terms of a sequence can never be coincident with the limit. Thus the implicit belief is slowly formed that a sequence of terms converging to a limit gets closer and closer, *but never actually gets there.*

Furthermore, if all the terms of a sequence have a certain property, it is natural to believe that the limit has the same property. Thus the sequence 0.9, 0.99, ... has terms all less than 1, so the limit "nought point nine recurring" must also be less than one... This leads to the mental image of a limiting object termed a *generic limit* in Tall (1986a). A generic limit need not be a limit in the mathematical sense, but it is the concept of the limit that the individual holds in his or her mind as a result of extrapolating the common properties of the terms of the sequence.

This phenomenon happens not just with sequences of numbers, but sequences of functions and other mathematical objects that share a common property. Historically this is enshrined in the "principle of continuity" of Leibniz:

In any supposed transition, ending in any terminus, it is permissible to institute a general reasoning, in which the final terminus may also be included.   (Leibniz in a letter to Bayle, January 1687.)

It arises even earlier in the work of Nicholas of Cusa (1401–1464) who regarded the circle as a polygon with an infinite number of sides, and inspired Kepler (1571–1630) to formulate a metaphysical "bridge of continuity" in which normal and limiting forms of a figure are characterized under a single definition. Thus Kepler (*Opera Omnia II*, page 595) saw no essential difference between a polygon and a circle, between an ellipse and a circle, between the finite and the infinite, and between an infinitesimal area and a line.

The generic extension principle arises time and again in history. For example, Cauchy's assertion that the limit of continuous functions is continuous and Peacock's "Principle of Algebraic Permanence", in which the properties of extended number systems, such as the real and complex numbers, were based on the principle that the any algebraic law which held in the smaller system also held in the extension. The latter held sway for some time in the nineteenth century until Hamilton invented (discovered?) the quaternions, an extension of the complex numbers whose multiplication fails to be commutative.

Obstacles arising from deeply held convictions about mathematics are rarely easy to erase from the mind. We all carry with us a mental rag-bag of such beliefs, many of which we suppress, but do not eliminate, when faced with the logic of mathematics. Often the only trace of such an obstacle is through a sense of unease when there is a logical deduction that does not "feel right". We view this as an instance of cognitive conflict between inconsistent portions of the individual's concept image.

1.7 GENERALIZATION AND ABSTRACTION

A common difficulty observed in students learning advanced mathematics is their complaint that the subject is "too abstract". What is the cognitive reason for their difficulty?

The terms "generalization" and "abstraction" are used in mathematics both to denote *processes* in which concepts are seen in a broader context and also the *products* of those processes. For instance, we *generalize* the solution of linear equations in two and three dimensions to n dimensions and we *abstract* from this context the notion of a vector space. In doing so two very different mental objects are produced: the *generalization* $\mathbb{R}^n$ and the *abstraction*, a vector space V over a field F. Practising mathematicians often regard a vector space as both an abstraction *and* a generalization of aspects of two dimensional space and it is therefore important to use the terms in a way which is consonant with their use in mathematics. But the mathematics educator must look at the cognitive processes which are involved, and here we see subtle differences between the two examples just given. The generalization $\mathbb{R}^n$ simply extends the chain of ideas from $\mathbb{R}^1$ to $\mathbb{R}^2$ to $\mathbb{R}^3$, and so on, which is *described* by applying the usual arithmetic processes to each coordinate. The abstraction V is a very different mental object, which is *defined* by a list of axioms. Whilst the former simply involves an *extension* of familiar processes, the latter requires a massive mental *reorganization*.

As Dreyfus will discuss in greater detail in chapter 2, the process of defining the abstract vector space must be followed by a sequence of theorems deducing the properties of a vector space which follow from the axioms. Cognitively this is not just a *deduction* process but a *construction* process in which the learner is building properties of the abstract object, for example, that the axioms guarantee the "usual" properties of addition of vectors and multiplication by scalars, that a linearly independent set of vectors will contain at most the same number of vectors as a spanning set, that a space with a finite spanning set has a precise "dimension" given in terms of an independent spanning set, or "basis", and so on. In this process of construction, existing examples of vector spaces (for example, $\mathbb{R}^2$, $\mathbb{R}^3$, etc.) act both as supports and potential conflict factors. They support because they suggest properties which are likely to hold, but they are potential sources of difficulty on the one hand because the learner is constrained to prove something that may seem "self-evident" from the examples and on the other because subtle properties common to all the examples may be believed to be "generically true" for the abstract concept. During this period there is a conflict between the properties of the examples which the learner *knows*, and the properties of the new abstract concept which must be *deduced* from the definition. A period of re-construction and consequent confusion is inevitable.

In Harel & Tall (to appear), we propose that a cognitive distinction be made between different types of generalization in accordance with the cognitive activities involved. We term an *expansive generalization* one which extends the student's existing cognitive structure without requiring changes in the current ideas. On the other hand, a generalization which requires reconstruction of the existing cognitive structure we call a *reconstructive generalization*. In this terminology we see that the general vector space $\mathbb{R}^n$ is, for most students, an expansive generalization, whilst the abstract vector space is both an abstraction and a *reconstructive generalization*.

We also note that it is possible for students in difficulties to operate in a third, subsequently disastrous, way which simply involves remembering the new ideas as an additional collection of information to be learned by rote and added to current knowledge without any attempt at integration with the old ideas. This we call a *disjunctive* generalization. It is a generalization in the sense that the student may now be able to operate on a broader range of examples, but it is likely to be of little lasting value to the student as it simply adds to the number of disconnected pieces of information in the student's mind without improving the student's grasp of the broader abstract implications.

The expansive generalization is a good teaching technique to adopt when it is necessary to be able to deal with a wider class of applications without having to go through too much stressful cognitive change. For instance, students who can carry out the process of solving simultaneous linear equations in two variables are usually able to generalize (expansively) to three, four, or more variables without difficulty (though the calculations may soon become tedious). Indeed, it is relatively straightforward to describe the process in general terms whilst referring to a specific set of equations in, say, three variables, x, y, z. ("Subtract suitable multiples of the first equation from the second and third to eliminate x, then eliminate y from the resulting equations, solve for z and substitute back to find y, then x.") The process is easier seen by enacting the solution than by describing it. Of course there may be exceptions (for instance "what to do when the first equation does not contain x"), but these may also be dealt with at the specific level. At risk of overusing an adjective we have already used before, we will regard this type of expansive approach as *generic*, in the sense that it describes the typical (general) procedure by referring to a *specific* case.

Such a generic approach is seen both an easy method of generalization because it applies a well-known process in a broader context and also as a first step towards formal abstraction as it does not involve a major cognitive reconstruction. Indeed, once the students have reflected on the general process and seen it as a conscious act of widening the applicability of a specific method, it may be viewed as a (relatively painless) form of abstraction which we term a *generic abstraction* (Harel & Tall, to appear). This furnishes an approach which is of particular value for students whose main interest is in applications rather than formal mathematics. It may also provide a suitable transition phase for students passing on to the formal abstraction, however, the latter will still require a cognitive re-organization, albeit one which is better prepared.

Dubinsky encourages students to write programs in a computer language where many of the constructs parallel the constructs of mathematical thinking: sets, sequences, ordered pairs, relations, functions, and so on. By writing computer code which specifies the procedure to carry out a function process, including an initial test to see if the input satisfies conditions which define the domain of the function, the student is required to think through the enactment of the function process. The act of programming is a generic process: it carries out what may be seen as a more general construct in particular cases and gives rise to a generic abstraction of the function concept. Given the theory just described, this suggests a further stage is necessary to pass from the generic example of programming, where the general is seen in the particular instances of functions programmed by the student, to the formal abstraction which requires a new level of abstract construction from the definition. Dubinsky formulates this transition within a Piagetian framework of *reflective abstraction*, in which processes are *encapsulated* as objects, so that the function *process* leads to the function *as a mental object*. This theory is further elaborated in chapters 7 and 15.

1.8 INTUITION AND RIGOUR

Mathematicians often regard the terms "intuition" and "rigour" as being mutually exclusive by suggesting that an "intuitive" explanation is one that necessarily lacks rigour. There is a grain of truth in this, for usually an intuition arrives whole in the mind and it may be difficult to separate its components into a logical deductive order. But the opposition between the two concepts is a false dichotomy as we shall soon see.

In a sense we have not one, but two brains. In attempting to assist patients who had serious epileptic fits, Sperry and his colleagues took the drastic action of partial or total severance of the corpus callosum that links the two hemispheres of the brain and found that each could essentially operate independently, though carrying out totally different functions:

Though predominantly mute and generally inferior in all performances involving language or linguistic or mathematical reasoning, the minor hemisphere is nevertheless clearly the superior cerebral member for certain types of tasks. If we remember that in the great majority of tests it is the disconnected left hemisphere that is superior and dominant, we can review quickly now some of the kinds of exceptional activities in which it is the minor hemisphere that excels. First, of course, as one would predict, these are all non-linguistic non-mathematical functions, largely as they involve the apprehension and processing of spatial patterns, relations and transformations. They seem to be holistic and unitary rather than analytic and fragmentary, and orientational more than focal, and to involve concrete perceptual insight rather than abstract, symbolic sequential reasoning.

(Sperry, 1974)

This evidence resonates strongly with the observation of the two different kinds of mathematical mind suggest at the turn of the century by Poincaré. However, subsequent research suggests that the brains of different individuals need not follow such a simplistic division of functions. Gazzigna (1985) sees brain activity as a collection of different modules functioning independently in parallel, with a control unit (usually in the left brain) making decisions based on the information provided by the various modules. Thus it would be incorrect to divide human activity simplistically into two different modes, just as it is inappropriate to consider just two contrasting types of mathematical mind. In particular we may envisage that the human mind immersed in logical thought may eventually develop intuitions that are themselves logically based. Poincaré, speaking of Hermite, said:

His eyes seem to shun contact with the world; it is not without, it is within he seeks the vision of truth.
    ... When one talked to M. Hermite, he never evoked a sensuous image, and yet you soon perceived that the most abstract entities were for him like living beings. He did not see them, but

he perceived that they are not an artificial assemblage and that they have some principle of internal unity.   (Poincaré, 1913, pp. 212, 220)

The conclusion is inescapable. Intuition is the product of the concept images of the individual. The more educated the individual in logical thinking, the more likely the individual's concept imagery will resonate with a logical response. This is evident in the growth of thinking of students, who pass from initial intuitions based on their pre-formal mathematics, to more refined formal intuitions as their experience grows:

We then have many kinds of intuition; first, the appeal to the senses and the imagination; next, generalization by induction, copied, so to speak, from the procedures of the experimental sciences; finally we have the intuition of pure number...   (Poincaré, 1913, p. 215.)

From a psychological viewpoint, Fischbein (1978) comes to similar conclusions, citing two different types of intuition:

Primary intuitions refer to those cognitive beliefs which develop themselves in human beings, in a natural way, before and independently of systematic instruction.
    Secondary intuitions are those which are developed as a result of systematic intellectual training ... In the same meaning, Felix Klein (1898) used the term "refined intuition": and F. Severi wrote about "second degree intuition" (1951).   (Fischbein, 1978, p. 161)

Thus aspects of logic too can be honed to become more "intuitive" to the mathematical mind. The development of this refined logical intuition should be one of the major aims of more advanced mathematical education.

2. THE GROWTH OF MATHEMATICAL KNOWLEDGE

As we have seen, the nature of mathematical thinking is inextricably interconnected with the cognitive processes that give rise to mathematical knowledge. We now focus on the full cycle of mathematical thinking to see mathematical proof as the final stage of this developmental process rather than just the formal framework of the completed knowledge structure.

2.1 THE FULL RANGE OF ADVANCED MATHEMATICAL THINKING

Mathematical proof, according to Hadamard (1945), is but the last, "precising" phase of mathematical thinking. Before a theorem can be conjectured, let alone proved, there is much work to be done in conceiving of what ideas will be fruitful and what relationships will be useful. Hadamard considers Poincaré's description of his own personal research activities and notes:

.. the very observations of Poincaré show us three kinds of inventive work essentially different if considered from our standpoint, *viz.,*

a. fully conscious work
b. illumination preceded by incubation
c. the quite peculiar process of the sleepless night.   (Hadamard, 1945, p. 35)

Here Poincaré reports the necessity of working hard at a new problem, then relaxing to allow the ideas to incubate in his subconscious, during which time he had a sleepless night thinking vigorously about new ideas until suddenly, some time later, a sudden illumination bursts into his consciousness with a solution. After a further time had elapsed, at his leisure, he was able to analyse what had happened and build up a formal justification of his theory in the final "precising" phase when the results of the illuminative break-through are subjected to the cold analysis of the light of day, refining the assumptions so that the deductions will stand analytic scrutiny.

What becomes apparent is that the initial phases of the creative cycle may rely in part on logic and deduction, but they also need flexible mental activity to produce mental resonances between previously unconnected concepts. According to Gazzigna's model of brain activity, they may occur as juxtapositions from different modules in the brain processing simultaneously. Part of the success of this phase of mathematical thinking seems to be due to working sufficiently hard on the problem to stimulate mental activity, and then relaxing to allow the processing to carry on subconsciously.

2.2 BUILDING AND TESTING THEORIES: SYNTHESIS AND ANALYSIS

Poincaré was at pains to show the complementary roles of synthesis and analysis in mathematical thinking. Synthesis begins with the conscious act of the initial phase to begin to put ideas together, followed by a more intuitive activity, in which subconscious interplay between concept images takes place, until a powerful resonance forces the newly linked concepts to erupt once more into consciousness. Analysis, on the other hand, is a much more cool and logical conscious activity which organizes the new ideas into logical form and refines them to give precise statements and deductions.

Teaching of younger children emphasizes the *synthesis* of knowledge, starting from simple concepts, building up from experience and examples to more general concepts. The emphasis at this level is now changing to include more problem solving and open-ended investigations. Teaching at university often emphasizes the other side of the coin: *analysis* of knowledge, beginning with general abstractions and forming chains of deduction from them which may be applied in a wide variety of specific contexts.

Working with much younger children, Dienes (1960) proposed a theory for building concepts from concrete examples, yet Dienes & Jeeves (1965) formulates a far more general *deep-end principle* in which "there is a preference for extrapolation by leaps and interpolation, rather than always by step-by-step". They respond to their own question "When is it possible to generalize from a simple case to a more general case and when is it better for them to particularize from a more complex case to the simple case?" with the remark that "this is not likely to be answered by a simple positive or negative statement". They suggest that it is more a question of "the optimum degree of complexity required to start with" – a response which is just as valid for teaching and learning at more advanced levels. It is likely to require synthesis of knowledge to build up theories cognitively as well as analysis of knowledge to give the total structure a logical coherence.

2.3 MATHEMATICAL PROOF

Viewed as a problem-solving activity, we see that proof is actually the final stage of activity in which ideas are made precise. Yet so much of the teaching in university level mathematics *begins* with proof. In his preface to *The Psychology of Learning Mathematics*, Skemp succinctly refers to this as showing the students the product of mathematical thought, instead of teaching them the process of mathematical thinking. The splendid tomes of Bourbaki are a monument to the intellect of the mathematical mind, and may be used to help the learner appreciate the formal structure of mathematics. But once again, Poincaré has pertinent observations to make:

To understand the demonstration of a theorem, is that to examine successively each of the syllogisms composing it and to ascertain its correctness, its conformity to the rules of the game? ... For some, yes; when they have done this, they will say: I understand. For the majority, no. Almost all are much more exacting they wish to know not merely whether all the syllogisms of a demonstrations are correct, but why they link together in this order rather than another. In so far as to them they seem engendered by caprice and not by an intelligence always conscious of the end to be attained, they do not believe that they understand.   (Poincaré, 1913, p. 431)

Perhaps you think I use too many comparisons; yet pardon still another. You have doubtless seen those delicate assemblages of silicious needles which form the skeleton of certain sponges. When the organic matter has disappeared, there remains only a frail and elegant lace-work. True, nothing is there except silica, but what is interesting is the form this silica has taken, and we could not understand it if we did not know the living sponge which has given it precisely this form. Thus it is that the old intuitive notions of our fathers, even when we have abandoned them, still imprint their form upon the logical constructions we have put in their place.   (*ibid*, p. 219)

Thus it is that so many mathematicians demand that a proof should not only be logical, but that there should be some over-riding principle that explains why the proof works. The proof of the four colour theorem, by exhaustion of all possible configurations using a computer search (Appel & Haken, 1976) *seems* logical, yet many professional mathematicians, though keen to see the theorem proved once and for all, are nevertheless sceptical that there may be some subtle flaw in the computer "proof", because there seems to be no rhyme or reason to illuminate why it works as it does.

Yet this principle is not always passed on to students. Sawyer (1987) reports how he tried to teach theorems in functional analysis by referring back to theorems in real variables that he expected his students to know, only to find that they had no recollection of them.

The reason for this was that in their university lectures they had been given formal lectures that had not conveyed any intuitive meaning; they had passed their examinations by last-minute revision and by rote.

He tells how he was shocked to learn of a lecturer who became stuck in the middle of a proof, turned his back on the class to draw a picture to aid him, then erased it and carried on with the formal proof without enlightening the class how he had used his intuition to rebuild it. He observes:

... to teach calculus well is a very demanding task. Three things have to be done: first to show by a drawing that some result is extremely plausible; second, to give counter-examples, which indicate the circumstances in which the conjecture would fail; third, to extract from these considerations a formal proof of the result.

These remarks do not apply only to lectures and books for undergraduate. Felix Klein pointed out that in papers for research journals the suppression of intuitive considerations was a common and highly undesirable practice.

(Sawyer, 1987)

Many mathematicians have learned to present their best face in public, showing their ideas in polished form and concealing the toil and false turnings that littered their growth. It is therefore essential to pose the following question:

> How it is possible to initiate students into the wider vision of the nature of mathematical thinking that includes the arduous growth of mathematical thinking *in a manner appropriate for a learner*?

3. CURRICULUM DESIGN IN ADVANCED MATHEMATICAL LEARNING

3.1 SEQUENCING THE LEARNING EXPERIENCE

During the difficult transition from pre-formal mathematics to a more formal understanding of mathematical processes there is a genuine need to help students gain insight into what is going on. A mathematician's logic may here fail him (or her) in designing a teaching schedule. A mathematician often takes a complex mathematical idea and "simplifies" it by breaking it into smaller components ready to teach each component in a logical sequence. From the expert's viewpoint the components may be seen as parts of a whole. But the student may see the pieces as they are presented, in isolation, like separate pieces of a jigsaw puzzle for which no total picture is available. In fact the scenario may be worse. As the student encounters each piece of the puzzle (s)he forms a personal concept image from the particular context which may be at variance with the formal idea. Thus, not only is no picture available for the puzzle, the pieces themselves may now have different shapes so that they no longer fit.

For example, a mathematical analysis of the notion of the derivative $f'(x)$ requires the notion of the limit of $(f(x+h)-f(x))/h$ as h tends to zero, so mathematically the derivative must be preceded by the discussion of the notion of a limit. To make the process mathematically easier the limit process is initially carried out with x fixed; only at a later stage is x allowed to vary to give the notion of a function. Thus the sequence suggested by a formal mathematical analysis is:

(1) notion of a limit,

(2) for fixed x, consider the limit of $(f(x+h)-f(x))/h$ as h tends to zero,

(3) call the limit $f'(x)$, then allow x to vary to give the derived function.

However, when the learner is at stage (1), the limit notion is mysterious because it seems "plucked out of the air", without any real reason. There are already cognitive obstacles here, as observed by Cornu (1983), and others. At stage (2) the limiting process introduces further obstacles (Tall & Vinner, 1981) which will be discussed in more detail in chapter 10. Nor is the passage from (2) to (3) as easy cognitively as it seems mathematically. Many students see (2) as a purely symbolic activity, and do not see the derivative f'(x) as a function, with a graph which is everywhere the gradient of the graph of f(x) (Tall, 1986).

The problem of curriculum development is therefore to present the student with contexts in which cognitive growth is possible, leading ultimately to meaningful mathematical thinking in which the formalism plays an appropriate part.

In analysis, for instance, one method which has proved successful might involve a more flexible approach that complements numerical and algebraic approaches to the derivative with a global, visual appreciation of the gradient of a graph generated on a computer.

In general it may be possible to use the complementary power of visualization to give a global gestalt for a mathematical concept, to show its strengths and weaknesses, its properties and non-properties, in a way that makes it a logical necessity to formulate the theory clearly. Visual ideas without links to the sequential processes of computation and proof are insights which lack mathematical fulfillment. On the other hand, logical sequential processes without a vision of the total picture, are blinkered and limiting. It is therefore a worthy goal to seek the fruitful interaction of these very different modes of thought.

3.2 PROBLEM-SOLVING

For many undergraduates, problem-solving means learning the contents of a set of lecture notes and applying this knowledge to specific problems clearly related to the material taught. For research mathematicians, problem-solving is a more creative activity, which includes the formulation of a likely conjecture, a sequence of activities testing, modifying and refining until it is possible to produce a formal proof of a well-specified theorem.

Polya (1945) suggested four phases as a framework for problem-solving:

- understand the problem,
- devise a plan,
- carry out the plan,
- look back at the work.

This framework has formed the backbone of many subsequent attempts at formulating problem-solving strategies, though Mason *et al* (1982) and Schoenfeld (1985) have seen the need to make the actual heuristics much more explicit and more appropriate for the learner. The idea of "devising a plan" is extremely daunting for the novice. More empathetic is the version suggested by Mason, who proposes three phases:

- entry,
- attack,
- review.

The entry phase covers the first two stages of Polya whilst attack and review correspond to Polya's third and fourth stage. In the *entry* phase the potential problem-solver gets acquainted with the problem-solving context – getting a sense of the problem by playing with the ideas, perhaps through simple specializations, moving to a position which attempts to specify clearly what is known and what is wanted, and considering carefully what can be introduced (notation, procedures of solution, etc.) that might take the problem-solver from what is known to what is wanted. Then a qualitative change occurs with a committed *attack* on the problem using the ideas that have been introduced. This may be successful, but it can more often lead to an impasse, a seeming dead-end from which the individual should review what has been done and return to the entry phase to consider a new attack. Once some kind of solution is achieved the mood changes yet again to one of sober *review* – checking the results to make sure no error has been made, reviewing what has been done to learn of strategies that may prove useful on other occasions and then being prepared to *extend* the problem to new levels of sophistication, re-starting the entry cycle at a more sophisticated level.

The author has had several years of experience teaching problem-solving within this framework. It has proved possible to get undergraduates to develop original ways of solving problems although the process requires longer initial periods for the students to reach a point of insight than may be apparent when giving the information in a lecture. However, the pay-off is in the way it can stimulate reflective thinking and develop an internal monitor within the student's mind to sense the progress and appropriate direction of the solution process.

3.3 PROOF

Students starting out in advanced mathematics have great difficulty with proof before they attain familiarity with the workings of the mathematical culture. In a questionnaire investigating which proof of the irrationality of $\sqrt{2}$ was more clear, students preferred a proof that showed that the square of any rational must have an even number of prime factors, and therefore such a square could not be 2 because the prime 2 occurs an odd number of times (namely once). They preferred this to the standard proof by contradiction and another more general demonstration taken out of Hardy's *Pure Mathematics*. This is despite the fact that this "proof" is not a formal proof at all, but a discursive explanation with examples demonstrating what form was taken by the square of a typical rational (Tall, 1979). Once more we see that a *generic* proof: explaining the general concept by considering a typical example, is an easier first step to understanding rather than the reconstructive leap to the general formalism.

Of course it is essential in advanced mathematics to take the step from (generic) explanation to formal proof. Some educators, such as Leron (1983ab, 1985ab), see their role as making the *structure* of proof more meaningful to students. His method is, essentially, to properly structure the proof, so that it is clear what is going on at any given time, and to make the proof as direct as possible. Thus contradiction proofs are re-written so that they are initially direct and constructive, with any contradiction being introduced as late as is practicable in the proof.

Others see their duty as the wider role of introducing students to the full range of mathematical thinking, including conjecture, positive verification through a convincing argument or refutation through a counter-example. Thus the Grenoble school (Legrand *et al*, 1984, 1988; Alibert, 1988) have introduced "scientific debate" into their courses in which a full lecture audience is invited to group together to think up likely theorems in the mathematical topic under consideration, and then to attempt to prove or disprove them. It is important that the teacher does not comment on the truth or falsity of the conjectures in the initial stages, so that the students are genuinely faced with the task of convincing their peers of the truth of their arguments. (See chapter 13.)

3.4 DIFFERENCES BETWEEN ELEMENTARY AND ADVANCED MATHEMATICAL THINKING

It is ironic that the National Curriculum in the UK and the NCTM standards in the USA for school mathematics advocate a level of open ended problem solving which is rarely specified in undergraduate courses at universities. The problem-solving procedures of entry, attack and review can *and are* being performed by younger children in such mathematical investigations. Thus many of the processes of advanced mathematical thinking are already found at a more elementary level. However, Mason *et al* (1982) describe the process of verification in *Thinking Mathematically* at three levels:

- convince yourself,
- convince a friend,
- convince an enemy.

Convincing oneself involves having an idea of why some statement might be true, but convincing a friend requires that the arguments be organized in a more coherent way. Convincing an enemy means that the argument must now be analysed and refined so that it will stand the test of criticism. This is the closest that *Thinking Mathematically* gets to the notion of proof. What is entirely absent is the notion of formal definitions and the logic of formal deductions from those definitions.

It may be hypothesized that mathematical thinking at every level can include the phases entry, attack and review, including a level of mathematical justification, but that elementary mathematical thinking lacks the process of formal abstraction and does not include the final "precising phase" in its most formal guise.

The move from elementary to advanced mathematical thinking involves a significant transition: that from *describing* to *defining*, from *convincing* to *proving* in a logical manner based on those definitions. This transition requires a cognitive reconstruction which is seen during the university students' initial struggle with formal abstractions as they tackle the first year of university. It is the transition from the *coherence* of elementary mathematics to the *consequence* of advanced mathematics, based on abstract entities which the individual must construct through deductions from formal definitions.

4. LOOKING AHEAD

It is a truism that we can only think with the cognitive structure that we have available to us. When we look at the psychology of advanced mathematical thinking, it is no wonder that we each find it easier to use our own knowledge structure to formulate our own theories.

As a mathematician entering mathematics education it is no surprise that the author first attempted to use catastrophe theory to describe the discontinuities in learning (Tall, 1977). Likewise those who begin mathematics education with a background of Piagetian theory are likely to attempt to explain things in these terms, those with experience in computer studies are more likely to use computer analogies, mathematicians are likely to attempt to use mathematical constructs, and so on. In trying to formulate helpful ways of looking at advanced mathematical thinking, it is important that we take a broad view and try to see the illumination that various theories can bring, the useful differences that arise and the common links that hold them together.

In the remainder of the first part of this book we consider the cognitive *processes* involved in advanced mathematical thinking and the two complementary attributes of the discipline: *creativity* in generating new ideas and the mathematician's notion of *proof* in convincing his peers of the truth of his assertions.

In the second part of the book we turn to cognitive theories that are proving of value in analysing the difficulties that students face and providing insights into the learning process that can be used in designing new ways of helping students construct mathematical ideas for themselves. First the differences between *concept definitions* and students' *concept images* are considered, then the nature of the mental objects which mathematicians construct: the *conceptual entities* that are the essence of advanced mathematics. This leads to the theory of *reflective abstraction* in which processes are encapsulated as mental objects which prove to be easier to manipulate at higher levels of abstraction.

In the third part of the text we review various advanced mathematical concepts from a cognitive viewpoint, showing the *cognitive obstacles* that can occur during their development and reporting empirical evidence on the success of instructional sequences designed from cognitive viewpoints. These involve the central ideas of *function, limit*, more advanced concepts of *analysis, infinity* and *proof*. We then move on to look at the new paradigm: the use of the computer and its cognitive effects in advanced mathematical thinking.

Finally, in chapter 15, as editor of the book, I take the opportunity of reflecting on the development of the theories of advanced mathematical thinking and its teaching and learning over the last decade and highlight the important themes which recur, the questions that have been posed and the partial answers that are beginning to show us the way ahead.

I : THE NATURE OF
ADVANCED MATHEMATICAL THINKING

What is it that is so difficult about Advanced Mathematical Thinking? In Part I of this book we have three chapters which consider the fundamental nature of advanced mathematical thinking to lay the groundwork for the cognitive theory and research reports to follow. First and foremost we acknowledge Advanced Mathematical Thinking in terms of creative process rather than just proof and deduction. In Chapter Two, Tommy Dreyfus recognizes that many of the processes of advanced mathematical thinking are also found in elementary mathematics. He considers the standard form of conveying information through lectures and reports consequent student difficulties in coping with anything which differs even marginally from what is taught. He focuses on the complexity of advanced concepts, the need to represent them and abstract their essential properties to control their complexity, and discusses the cognitive difficulties in carrying out these processes. In Chapter Three, Gontran Ervynck considers the enigmatic nature of the creative process in mathematics, which is the focus of the research process yet plays so little part in student development. In Chapter Four, Gila Hanna analyses the nature of mathematical proof from a philosophical and pragmatic viewpoint to show that it is dependent on context and beliefs rather than an immutable standard shared by all mathematicians.

In these three chapters we therefore take a critical look at advanced mathematical thinking as part of the living process of human thought rather than the immutable final product of logical deduction.

ADVANCED MATHEMATICAL THINKING PROCESSES

TOMMY DREYFUS

Understanding, more than knowing or being skilled, has always been considered an important goal by mathematics teachers. Understanding, as it happens, is a process occurring in the student's mind; it may be quick, an "Aha-Erlebnis", a click of the mind; more often, it is based upon a long sequence of learning activities during which a great variety of mental processes occur and interact. Therefore, what it means to come to understand a mathematical notion or concept is extremely difficult to analyze. Researchers in psychology (e.g., Brown, Bransford, Ferrara, & Campione, 1983) have been asking themselves what "understanding" means, in particular what are its components, what mental processes may intervene and combine together to form that meta-process of understanding. Researchers in mathematics education, in particular, have become conscious of the importance of the component processes for understanding advanced mathematics and their interactions.

Why would researchers be interested in the processes involved in learning advanced mathematics? One reason is to gain basic theoretical knowledge about what is going on in the student's mind. There certainly is some intrinsic interest in this fundamental question. But there are also very important applied aspects to this strand of research, and these concern all teachers of advanced mathematics. The processes the teacher hopes to provoke in the student do not happen by themselves nor, if they happen, are they necessarily conscious on the students part. It is not sufficient, for example, to define and exemplify an abstract concept such as vector space. Students must then construct the properties of such a concept through deductions from the definition. They may involve being through activities that promote abstraction on their part and it has to be brought to their attention that this is what is being done, that this is the aim of the exercise. In this chapter, processes, among them abstracting, are analyzed and discussed with the aim of making teachers of advanced mathematics more conscious of what is going on during these processes. Hopefully, this will help teachers introduce such action explicitly in their classrooms. Recently some controlled trials with similar aims have been made, and they have met with some success. A conscious process approach to abstracting has been described by Mason (1989) and experiences with making student teachers reflect upon their mathematical activity have been reported by Southwell (1988).

Reflection about one's mathematical experience is of particular importance in the solution of non-trivial problems (as opposed to standard exercises). And it is in this connection that the importance of processes has first been realized by mathematics educators (Schoenfeld, 1985). Reflection about one's mathematical experience is an important aspect of meta-cognizing, another meta-process. Such reflection is a characteristic of advanced mathematical thinking. We would not usually expect an elementary math student to stop, after having solved a problem, and think or recount how he went about solving this problem. We would, however, definitely like to see much more of this in our advanced students and, in particular, in our high school teachers.

There is no sharp distinction between many of the processes of elementary and advanced mathematical thinking, even though advanced mathematics is more focussed on the abstractions of definition and deduction. Many of the processes to be considered in this chapter are present already in children thinking about elementary mathematics concepts, say number or place value. They are not *exclusively* used in advanced mathematics, nor, indeed, are they exclusively used in mathematics. Abstractions are made in physics, representations are used in psychology, analysis is used in economics and visualization in art. Here, however, we will describe the processes as they are relevant for advanced mathematical thinking, in particular focussing on those processes whose characteristics make the mathematical thinking advanced.

It is possible to think about advanced mathematical topics in an elementary way (e.g., many standard exercises on rings or groups can be answered by just plugging in the right numbers), and there is rather advanced thinking about elementary topics (look at some of the problems in mathematics olympiads). One distinctive feature between advanced and elementary thinking is complexity and how it is dealt with. Advanced concepts, such as rings or Lie groups, are likely to be very complex. The distinction is in how this complexity is managed. The powerful processes are those that allow one to do this, in particular abstracting and representing. By means of abstracting and representing, one can move from one level of detail to another and thus manage the complexity.

The processes to be discussed in this chapter are mathematical and psychological ones, and in many cases they are both; in fact, the mathematical and the psychological aspects of a process can rarely be separated. For example, when you build a graph of a function, you are executing a mathematical process, following certain rules which can be stated in mathematical language; at the same time, however, you are very likely generating a visual mental image of that graph; in other words, you are visualizing the function in a way that can later help you reason about the function. The mental and the mathematical images are closely linked here. Neither can arise without the other, and they are in fact generated by the very same process; they are, respectively the mathematical and the psychological aspects of this process. A similar linkage between mathematics and psychology exists with respect to the other processes of advanced mathematical thinking. In fact, it is precisely this linkage which makes the processes interesting and relevant for understanding learning and thinking in advanced mathematics.

1. ADVANCED MATHEMATICAL THINKING AS PROCESS

The typical mathematics course at, say, first year university level has a neatly defined syllabus, which tells the instructor what material he is supposed to cover by the end of the term. Whether this is a course in calculus, algebra, finite mathematics, numerical methods or other, for the instructor the content to be taught is a well known, unassailable, accepted segment of mathematical knowledge; although (s)he will probably think of several possibilities to organize this material into a logically clean structure, each of these structures will basically consist of a number of theorems, to be proved, and a number of applications of these theorems to topics in mathematics and beyond. The instructor will probably distribute these into as many class periods as are available and lecture during a considerable part of these class periods, making extensive use of the strikingly convenient formalisms of the specific domain of mathematics concerned. In so doing, a very important aspect of the mathematics which is being taught is presented to the students, namely the finished and polished product into which that well known, unassailable, fully accepted segment of mathematics has grown.

Our instructor presumably knows very well that mathematics is not being created in final and polished form, but through trial and error, through partially correct (and partially wrong) statements, through intuitive formulations in which loose terms and imprecisions have intentionally been introduced, through drawings that try to visually present parts of the mathematical structures being thought about, through dynamic changes being made to these drawings, etc., etc.. But the fact that our instructor knows about these other aspects of mathematics, indeed, is very likely to experience them daily in his or her research work, does not usually prevent him or her from almost exclusively teaching the one very convenient and important aspect of mathematics which has been described above, namely the polished formalism, which so often follows the sequence theorem–proof–application. This manner of teaching has several advantages: for example, it allows for a well-planned structure of the course, as well as for predictable progress through the material, and thus for a fairly certain guarantee that most of the material in the syllabus can be covered. Unfortunately, it also has at least one very serious disadvantage: it is inflexible in terms of adaptability to the students. It may work rather well for students who major in mathematics and who, from some exceptional teacher or on the basis of their own talent and investigative nature, have already had the opportunity to acquire a mathematical attitude. But as is shown, for instance, by the present calculus crisis, it does not work for the vast majority of students, those majoring in science, engineering, medicine or the liberal arts and taking mathematics as a required service subject.

What these students learn, and what they don't learn, is illustrated very well by the results of a recent study of students who had passed a traditional first quarter calculus course at Tennessee Technological University (Selden, Mason & Selden, 1989). These were well prepared students who had been taught in small groups by experienced teachers and teaching assistants, obtaining at least a C grade. The students were presented with five moderately difficult problems that could easily be solved with the techniques at the students' disposal. These problems were formulated in a manner that was somewhat non-standard, for example:

Find at least one solution to the equation $4x^3-x^4=30$ or explain why no such solution exists.

The function $f(x)=4x^3-x^4$ has a maximum value of 27 and thus no solution to the given equation exists. Although the students in the study were perfectly capable to carry out the required function discussion, they could not answer the question as given. The situation was similar with respect to the other four problems. In fact, the authors state that *not one student got an entire problem correct. Most couldn't do anything.*

This is not an exceptional finding, nor is it limited to average university freshman. Davis (1988), in discussing a class of *unquestionably excellent* high school seniors, came to the conclusion that *when one looks carefully at how these "apparently successful" students deal with mathematical problems, one finds that they hold many grotesque misconceptions about mathematics* and that the following reasons contribute to their success in passing tests: *Most mathematics instruction, from elementary school through college courses, teaches what might be called rituals: "do this, then do this, then do this ..."* and *Teachers ... will typically accept the correctly-performed ritual as enough success for the time being.*

In other words, what most students learn in their mathematics courses is, to carry out a large number of standardized procedures, cast in precisely defined formalisms, for obtaining answers to clearly delimited classes of exercise questions. They thus acquire the capability to perform, albeit much slower, the kind of operation which a computer can perform by means of a suitable program such as *Mathematica*. They end up with a considerable amount of mathematical knowledge but without the working methodology of the mathematician, that is they lack the know-how that allows them to use their knowledge in a flexible manner to solve problems of a type unknown to them. They are examples of the phenomenon described in the previous chapter: they have been taught the products of the activity of scores of mathematicians in their final form, but they have not gained insight into the processes that have led mathematicians to create these products.

While there is no need to form every student of science or engineering into a full-fledged mathematician, most teachers of calculus would like their graduates to be able to answer questions such as

What conditions are sufficient to ensure that the function $f(x)=ax^3+bx^2+cx+d$ is increasing at $x=0$?

and to immediately start searching for a mistake when they see a result that obviously has the wrong sign such as in

$$\int_{-1}^{+1} \frac{1}{x^2}\, dx = \left[-\frac{1}{x}\right]_{-1}^{+1} = -2.$$

Moreover, one would expect that they realize rather quickly that

$$\int_a^b g(x)\, dx = \int_{a+k}^{b+k} g(x-k)\, dx$$

for any continuous function g and that they have little difficulty in determining which of the statements (i) $f'(a)=-f'(a)$, (ii) $f'(-a)=f'(a)$, (iii) $f'(-a)=-f'(a)$, (iv) $f'(-a)=-f'(-a)$ are true for any odd differentiable function f. Experience shows, however, that such tasks are difficult for calculus students, and that a similar situation pertains to students of other courses such as pre-calculus mathematics, linear algebra and differential equations.

The discrepancy between teacher expectations and student performance on such tasks occurs because we, as teachers, often do not realize how much of our experience with mathematical processes and concepts we use. Take, for example, the equality

$$\int_a^b g(x)\,dx = \int_{a+k}^{b+k} g(x-k)\,dx;$$

here one needs to deal with the function g as an object that is operated on in two ways, namely integrated from a to b to give a number as well as translated to the right and then integrated over a translated interval; one possibility is to visualize the translations of the function and the interval and to compare them. Although all of this may hardly take more than a few seconds for the expert, his mental processing has probably included components of representing (the function, maybe graphically), transforming (by translation), visualizing (the function, the translation, the area under the graph), checking (that the two translations go in the same direction), and deducing (that the resulting numbers are equal). Possibly processes of specializing (e.g., to positive functions only) and then generalizing again have also been involved. Moreover, all of this was most certainly based on extensive prior processing of functions and integrals which included repeated phases of generalizing, abstracting and formalizing, that allow the expert to view functions and integrals as objects but may not be available to the student. The message that I am trying to convey here is that advanced mathematical thinking, as in the expert's treatment of the equality of these two integrals, is an extremely complex process, in which a large number of component processes interact in intricate ways.

One place to look for ideas on how to find ways to improve students' understandings is the mind of the working mathematician. Not much has been written on how mathematicians actually work; certainly the deepest and most elaborate document on this topic is Hadamard's book (1945), which refers back extensively to Poincaré's ideas and was the source of inspiration for much of the discussion in the previous chapter. Hadamard explicitly stresses the importance of informal reasoning, of thinking in the absence of words, of visual imagery, of mental images (which cannot necessarily be expressed in words, at least not when they first occur), and of playing around with ideas, for instance by repeatedly trying to fit different elements together much like in a puzzle. This experimental aspect of mathematics, as well as the visual means being used, have received added importance in recent years due to the technological progress that has been made in computer graphics. Investigations have become possible which are in principle qualitative and visual. The prime example for this is research on non-linear systems, chaos and fractals. According to Peitgen & Jürgens (1989), the fundamental mathematical developments in this area were made possible only through computer-graphical experiments.

Hoffman (1989) has proposed a philosophy of mathematics education based on the simple recognition that mathematics is a human activity, useful in the real world; on this basis he requires that we should transmit to our students a picture of mathematics as a science which incorporates observation, experiment and discovery.

Several projects have been carried out in recent years following basically this philosophy. For example, Breuer, Gal-Ezer & Zwas (1990) proposed to teach numerical methods at calculus level in a computational laboratory and Ruthven (1989) has reported how the use of graphic calculators has led students to use a graphic-trial approach and a numeric-trial approach in parallel to an analytic-construction approach to pre-calculus problems. Many (though not all) similar projects make extensive use of computers as experimental tools. Computers can serve as heuristic tools for the mathematician and the mathematics student in much the same way as a microscope serves the biologist: if the tool is directed onto interesting phenomena and correctly focussed, it may show an unexpected picture, often a visual one, of the phenomenon under study, and thus lead to new ideas, to the recognition of heretofore unknown relationships. In the case of the researcher, these ideas and relationships may be expected to be original; in the case of the student, they were very likely known to many other people before but they are new to this particular individual.

By using computer learning environments many usually implicit relationships, for instance between different representations for the same concept, may be made explicit. This explicitness contributes to students' recognition of such relationships and to the emergence of related ideas, in brief to their formation of concepts. This process thus closely parallels the recognition of relationships and the emergence of ideas in the research process. Admittedly, there are clear differences between the research process and the learning process; for instance, in the learning process the material to be learned is presented in a manner judged by experts to be digestible; and the average learner should be expected to be considerably less talented for mathematics than the average researcher; but the point here is to stress the very important similarities between the learning process and the research process; namely that in both cases the individual has to mentally manipulate, investigate and find out about objects, about which his knowledge is very partial and fragmented. Thus, just as the research process is extraordinarily complex, so is the corresponding learning process. It contains the gist of what advanced mathematical thinking is all about. It is likely to comprise, at any stage and in tight interaction, several of the processes mentioned in the discussion of the translated integral, e.g. representing, visualizing, generalizing, as well as others such as classifying, conjecturing, inducing, analyzing, synthesizing, abstracting or formalizing. In other words, advanced mathematical thinking consists of a large array of interacting component processes. It is important for the teacher of mathematics to be conscious of these processes in order to comprehend some of the difficulties which their students face.

2. PROCESSES INVOLVED IN REPRESENTATION

2.1 THE PROCESS OF REPRESENTING

Representations have a very important function in mathematics: If we want to talk about the group of permutations of n objects, for instance, it will often be convenient to call it the symmetric group of degree n and denote it by S_n. The notation S_n is a sign that refers to, and thus represents, or symbolizes, the group in question; it is a symbolic representation of the group. Such symbols are absolutely indispensable in modern mathematics, but there is also some danger associated with them. As has been stressed by Olson & Campbell (in press), symbols involve relations between signs and meanings; they serve to make a person's implicit knowledge – the meaning – explicit in terms of symbols. There must be some meaning associated with a notion before a symbol for that notion can possibly be of any use; in the educational discourse of teaching mathematics, this is too often overlooked leading to the well known phenomenon of "symbol pushing". The role of symbols is discussed in greater detail in chapter 6 by Harel and Kaput.

Another meaning of representations is even more central for learning and thinking in mathematics. When we talk or think about a group, an integral, an approximation, about any mathematical object or process at all, each one of us relates to something we have in mind – a mental representation of the object or process under consideration. Although most mathematicians can be expected to come up with roughly equivalent definitions of, say, a function, their respective mental representations of the notion may be vastly different. Have you ever asked mathematicians working in different areas what comes to their mind when they think about functions? When you also ask mathematics teachers and students, these differences become not only more pronounced but also much more important. For example, a student's notion of a function may be limited to processes (of computation or mapping), whereas the teacher teaching indefinite integrals may think of the function in the integral as an object to be transformed. Such discrepancies easily lead to situations where students are unable to understand their teachers.

To represent a concept, then, means to generate an instance, specimen, example, image of it. But this short description is insufficient for us, because it does not specify whether the generated instance is symbolic or mental, nor does it indicate what "generate" means in terms of the processes by which mental representations come into existence and how they develop. A symbolic representation is externally written or spoken, usually with the aim of making communication about the concept easier. A mental representation, on the other hand, refers to internal schemata or frames of reference which a person uses to interact with the external world. It is what occurs in the mind when thinking of that particular part of the external world and may differ from person to person. In case of the symmetric group, one person's mental representation may consist of nothing but the symbol S_n, another may think of a set of colored cubes which are being permuted, a third may see "in the mind's eye" symbols like (1 3 5)(2 4 6 7) which may or may not have an associated meaning, and still another person may conceive of the group by way of its irreducible representations. Another example is that of vector space. When I think of a vector space, I may "see" arrows (before my mind's eye), and I may be able to think in terms of these arrows when dealing with bases, transformations etc. Others may evoke n-tuples of numbers or abstract symbols which satisfy the axioms.

Visualization plays an essential role in the work of many eminent mathematicians. For instance, Einstein wrote to Hadamard:

Words and language, written or oral, seem not to play any role in my thinking. The psychological constructs which are the elements of thought are certain signs or pictures, more or less clear, which can be reproduced and combined at liberty.   (Hadamard, 1945, p. 82)

Visualizing is one process by which mental representations can come into being. A more general description of how mental representations of mathematical concepts may be generated has been proposed by Kaput (1987b); according to his theory, the act of generating a mental representation, relies on representation systems, i.e. concrete, external artefacts, which can be materially realized. In the case of functions, graphs are one such artefact, algebraic formulas are another, arrow diagrams and value tables still other ones. Mental representations are created in the mind on the basis of these concrete representation systems. A person may thus create a single or several competing mental representations for the same mathematical concept. The topic of students' mental images of various mathematical concepts has already been mentioned in the opening chapter of this volume and will be discussed further by Vinner in chapter 5.

To be successful in mathematics, it is desirable to have rich mental representations of concepts. A representation is rich if it contains many linked aspects of that concept. A representation is poor if it has too few elements to allow for flexibility in problem solving. Such inflexibility we often observe in our students: The slightest change in the structure of a problem, or even in its formulation, may completely block them (see, e.g., the study by Selden, Mason & Selden described in the previous section). Poor mental images of the function concept, for instance, are typical among beginning college students, who think only in terms of formulas when dealing with functions, even if they are able to recite a more general set-theoretic definition (see Eisenberg, chapter 9).

Several competing mental representations of a concept may coexist in somebody's mind, and be taken advantage of: different ones may be called up for considering different mathematical situations. However, different mental representations may also enter into conflict such as, for example, in a calculus student described by Schoenfeld, Smith & Arcavi (in press), who simultaneously held four competing and conflicting interpretations of the y-intercept of a straight line. In more favorable cases, several mental representations for the same concept may complement each other and eventually may be integrated into a single representation of that concept. This process of integration is related to abstraction and is further discussed below. As a result of this process, one has available what is best described as multiple-linked representations, a state that allows one to use several of them simultaneously, and efficiently switch between them at appropriate moments as required by the problem or situation one thinks about.

2.2 SWITCHING REPRESENTATIONS AND TRANSLATING

Although it is important to have many representations of a concept, their existence by itself is not sufficient to allow flexible use of the concept in problem solving. One does not get the support that is needed to successfully manage the information used in solving a problem, unless the various representations are correctly and strongly linked. One needs the possibility to switch from one representation to another one, whenever the other one is more efficient for the next step one wants to take. The process of switching representations is thus closely associated with that of representing. Switching must always be carried out between existing representations. In our context, it means going over from one representation of a mathematical concept to another one. And again, functions are probably the clearest example. A function is an abstract concept with which we usually work in one of several representations, or preferably in several representations at once; often these include the graphical and the algebraic representation. Teaching and learning this process of switching is not easy because the structure is a very complex one; think, for example, of a trigonometric function, which already has the properties of frequency, amplitude and phase; now consider the algebraic formula for this function, its graph, and a table containing the values of the function at special points such as extrema and zeros; moreover establish the links between these three representations, e.g. clarify to yourself where some of the points in the table of values appear on the graph or how the value of the phase determines the position of the graph; this is already a lot of information to be dealt with, especially for students who lack extensive experience; but all of this information may only be the starting point for a problem such as the behavior of the trigonometric function under a shift or stretch transformation, i.e. a change of frequency or phase. As a consequence, students very often limit themselves to working in a single representation; for example, even when they are required to draw a sketch, say before integrating an absolute value function, they often ignore their own sketch and thus fail to solve the problem correctly (Mundy, 1984). One possible approach is to systematically use several representations in teaching and to stress the process of switching representations from the beginning. Computer environments have been successfully used to achieve this in curricula for functions (Schwarz, Dreyfus & Bruckheimer, 1990), calculus (Tall, 1986a, 1986b), and differential equations (Artigue, 1987). These will be further discussed in chapter 11 by Artigue and chapter 14 by Tall & Dubinsky. The way in which multiple-linked representations may be treated is well illustrated by the functions curriculum. This introductory curriculum is based on a computer micro-world, which has been specifically designed to encourage switching representations. Students are asked to solve open ended problems such as maximizing, to a given accuracy, the volume of an open box constructed from a given sheet of cardboard (before they learn any calculus!). To successfully solve this problem they have to use at least two representations, and they need to transfer information obtained in one representation in order to use it in another one. A large majority of students in the study have been found to transfer information between representations and to successfully use the transferred information for solving problems. It has been concluded that, for these students, functional representations are symbols in the sense described above, namely signs with associated meaning; moreover, that meaning was common to several representations of the function; in other words, these students developed a satisfactory function concept incorporating three representations, between which they were able to switch during problems solving processes. (Schwarz & Dreyfus, 1991).

A process which is closely connected to switching representations is translating. One meaning of translating which is relevant for advanced mathematical thinking is going over from one formulation of a mathematical statement or problem to another one. Applied problems are a case in point. A second order linear differential equation with constant coefficients may be presented as an oscillation problem, possibly with friction; its solution may then be discussed in terms of permanent and transient states. From the present point of view, this constitutes an additional representation, and one that introduces considerable additional difficulties for the beginning student; at least my students are very apprehensive about "applied problems" in examinations. This can be easily explained in the light of the above discussion. Not only does the student need to understand the context of the applied problems, e.g. an electric circuit, but more importantly, he needs to establish a close and clear correspondence between quantities referred to in terms of electric circuits and quantities referred to in terms of differential equations. This correspondence may be obvious to the teacher, but for the student the construction of the appropriate mental schemata is a difficult task which needs to be supported by explicit teacher action.

In this subsection, the distinction, made above, between mental and concrete and symbolic representations has been blurred. To some extent, this is necessary when learning processes are considered, because mental representations arise from concrete ones. The section on representing will now be concluded with a process of concrete representing that has particular importance in applied mathematics and takes on ever greater weight in university, and more recently also in school curricula, namely modelling.

2.3 MODELLING

Typically, the term modelling refers to finding a mathematical representation for a non-mathematical object or process. In this case, it means constructing a mathematical structure or theory which incorporates essential features of the object, system or process to be described. This structure or theory, the model, can then be used in order to study the behavior of the object or process being modelled. For example, the Schrödinger equation models the behavior of certain physical systems which obey the rules of quantum mechanics; or a crystallographic group models the symmetry properties of a chemical compound. A mathematical model thus has the status of a representation of a (physical) situation; but for the person thinking about, say the symmetry properties of a silicate crystal, this is not enough; that person also needs a mental representation of the silicate's symmetry group. This leads to an interesting connection between a model and a mental representation. The process of representing is, to some extent, analogous to the modelling process, but on another level. In modelling the situation or system is physical and the model is mathematical; in representing the object to be represented is the mathematical structure, and the model is a mental structure. Thus the mental representation is related to the mathematical model as the mathematical model is related to the physical system. Each is a partial rendering of the other. Each reflects some (but not all) properties of the other. And each enhances one's capacity to mentally manipulate the system under consideration.

3. PROCESSES INVOLVED IN ABSTRACTION

Many of the processes mentioned in this book occur at any level of mathematical thinking: Certainly, even small children create mental representations of anything they think about, and particularly of mathematical objects of thought, such as numbers or triangles. Starting no later than in elementary school, children also work with these objects, especially numbers, in different representations. Other processes, however, take on added importance as students' mathematical experience and abilities develop and as the mathematical contents they deal with become more advanced; the most important among these advanced processes is abstracting. If a student develops the ability to consciously make abstractions from mathematical situations, he has achieved an advanced level of mathematical thinking. Achieving this capability to abstract may well be the single most important goal of advanced mathematical education.

Two processes, in addition to representing, form a prerequisite basis to abstracting: *generalizing* and *synthesizing*; we briefly discuss these first.

3.1 GENERALIZING

To generalize is to derive or induce from particulars, to identify commonalities, to expand domains of validity. A student may know from experience that a linear equation in one variable has one solution, and that "most" systems of two (three) linear equations in two (three) variables have a unique solution. (S)he may then generalize this to systems of n linear equations in n variables. More importantly, with appropriate guidance, (s)he may be led to examine the meaning of "most" for $n=2$ and $n=3$ in the above statement, formulate it as an appropriate condition, and generalize also that condition to $n>3$. In this process, one needs to make the transition from the particular cases of $n=1, 2, 3$ to general n, one needs to identify what the conditions for $n=2$ and $n=3$ have in common, and to conjecture and then establish that the domain of validity of the conclusion "there is a unique solution" can be extended to general n.

The generalization in the previous example is important in that it establishes a result for a large class of cases – all systems of n independent linear equations in n variables. It is, however, limited to establishing analogies between the concrete cases of $n=2$ and $n=3$, and extending them to the case of n equations in n variables, which may be less concrete but presents no essentially new features. In particular, the general case does not require the formation of any mathematical concepts that were not present for $n=3$. Other generalizations do include the need for such concept formation. An example is the transition from convergence of a sequence of numbers to convergence of a sequence of functions, which gives rise to the need for a topology on the space of functions. The cognitive requirements in the process of generalization are thus increased considerably, and for the specific case of convergence of functions the degree of difficulty of these requirements is well documented by several decades of discussions between Cauchy, Fourier and Abel at the beginning of the 19th century (Lakatos, 1978). Students must thus be expected to have a hard time with such generalizations, and experience confirms this. It must be pointed out, however, that even in this case, the generalization takes place with respect to given (mathematical) objects, equations in the first case, numbers and functions in the second. The presence of these objects is helpful to the student because it leaves him on (hopefully) well known, solid ground while trying to grapple with the added generality of the situation.

3.2 SYNTHESIZING

To synthesize means to combine or compose parts in such a way that they form a whole, an entity. This whole then often amounts to more than the sum of its parts. For example, in linear algebra, students usually learn quite a number of isolated facts about orthogonalization of vectors, diagonalization of matrices, transformation of bases, solution of systems of linear equations, etc. Later in the learning process, all these previously unrelated facts hopefully merge into a single picture, within which they are all comprised and interrelated. This process of merging into a single picture is a synthesis. Thurston (1990) has recognized that, due in part to this possibility of synthesis, mathematics is tremendously compressible. He has also noted, that while the insight that goes with this compression is one of the real joys of mathematics, this process is irreversible; therefore, it is very hard for the mathematician to put himself in the frame of mind of the student who has not yet achieved this synthesis, and to see not only how much detail is involved in learning even simple concepts and operations but how much detailed work with these concepts and operations is needed to be able to start synthesizing.

Classroom practice often does not put enough stress on this process of synthesis. While the details are explained at length by the teacher and exercised by the student, few if any activities are designed to lead the student to synthesize different aspects of a concept, and even less different concepts within a domain or even different domains. Obviously, the good teacher does his part of summarizing and this often includes some synthesis. He does state, for the students' benefit, what the connections, relationships, etc. are. But the fact that it is done by the teacher rather than in a student activity conveys to the students that this is what the mathematicians see, and is of no direct relevance to the problems the student has to solve. These problems are standard exercises, which do not require synthesis. Consequence: I do not need it for the exam, why should I bother? As we saw earlier in the chapter, non-standard questions, even if almost trivial, but requiring some amount of flexibility of thinking and synthesis, are usually out of reach, at least for the average student (Selden, Mason & Selden, 1989). Students, specifically high school students who do well in mathematics, believe that solving a mathematics problem should typically take one minute, and never more than ten; they also think that memorizing is extremely important for success in mathematics and that there is little relationship between the different mathematics courses (algebra, geometry, trigonometry) which they have taken (Schoenfeld, 1989). Again, even if synthesis may be in the teacher's mind, it is sorely lacking from the student's.

3.3 ABSTRACTING

In the transition from the concrete vector space $\mathbb{R}^3$ to the notion of an abstract vector space, the relationships between the vectors become the focus of attention, whereas their specific realization as triples of numbers is dropped. In order to make this transition, one needs to be able to conceive of the object "vector" purely in terms of its relationships to other similar or different objects (vectors or scalars), and accept that the object itself is not further specified by any intrinsic properties. Considering only these relationships, enables one to draw conclusions from them which will be generally valid, independently of the specific intrinsic properties of the vectors. In this manner, much of the power of mathematics derives from abstraction.

The process of abstraction is thus intimately linked to generalization. One of the main incentives for abstraction is the general nature of the results that can be obtained. Another incentive is the achievement of synthesis. Groups show to the student that it is possible to describe in a unifying manner a vast number of situations that have heretofore been considered separately and independently of each other. But neither generalizing nor synthesizing make the same heavy cognitive demands on students as abstraction. As we saw in chapter 1, generalization usually involves an expansion of the individual's knowledge structure whilst abstraction is likely to involve a mental re-construction. In the transition, say, from real to complex numbers, we achieve generalization by not insisting any more on order but we continue working with objects that are represented explicitly using numbers that we can add and multiply in a familiar way. Similarly, a student may well learn about the connections between matrix calculus and planar or spatial symmetry transformations such as principal axis transformations or crystallographic point groups without forgoing their explicit concrete realization. Abstraction, however, requires giving up exactly this explicitness: the student is required to focus on the relationships that exist between numbers in order to be able to grasp what a field is, rather than on the numbers themselves, and similarly for other notions such as function, group and vector space.

Abstraction thus contains the potential for both generalization and synthesis; vice versa, it gets its purpose mainly from this potential of generalization and synthesis. The nature of the mental process of abstracting is, however, very much different from that of generalizing and from that of synthesizing. Abstracting is first and foremost a *constructive* process – the building of mental structures from mathematical structures, i.e. from properties of and relationships between mathematical objects. This process is dependent on the isolation of appropriate properties and relationships. It requires the ability to shift attention from the objects themselves to the structure of their properties and relationships. Such constructive mental activity on the part of a student is heavily dependent on the student's attention being focussed on those structures which are to form part of the abstract concept, and drawn away from those which are irrelevant in the intended context; the structure becomes important, while irrelevant details are being omitted thus reducing the complexity of the situation. The role of mathematical and mental structure in abstraction has been examined by Thompson (1985a) and Harel & Tall (in press). The cognitive aspects of focussing and shifting attention during the process of abstraction have been investigated by Dörfler (1988) and Mason (1989). Abstraction will be further discussed by Dubinsky in chapter 7 from a Piagetian viewpoint. We here only raise a few points which may serve as background for the coming chapters.

It is an open didactical problem, whether one should lead students to abstract from many cases or from a single one. Schoenfeld, Smith & Arcavi give a very detailed description of how one above average student constructed her understanding of y-intercept (of a straight line). They observed her give four different, inconsistent interpretations of y-intercept depending on the context she had to deal with; e.g., she interpreted y-intercept differently according to whether the line was given by two points on the same or on different sides of the y-axis. It took seven hours of work over several weeks for the student to decontextualize the notion and achieve an abstract unified concept of y-intercept. Given the instability in the student's interpretations, it must be considered unlikely that a single example and an explicit formal definition would have helped her avoid later misinterpretations. More generally, having several examples, e.g. of concrete groups, will enable students to identify commonalities; this is one way for the teacher to focus students' attention on those properties and relationships which are important for the intended abstraction. And this way of focussing attention may work well, if the amount of information in the detailed description of the internal structure of the examples is limited. If, however, the examples are too complex, i.e. if they have many properties that are to be ignored in the process of abstracting, it may be difficult to achieve such focussing. Therefore, it is sometimes better to abstract from a single case, combined with a definition of the abstract concept. This single case then needs to be chosen so that the intended properties and relationships take some evidence, e.g. by being useful in an activity the students engage in. Students' experience with making abstractions is also likely to be a factor here: Grade schoolers who should learn about place value (a very abstract and difficult concept) are unlikely to have much use for a definition; mathematics students who already have a good grasp of what vector spaces and groups are, may not need dozens of examples of rings before being able to digest a definition. The question is thus one of finding the good measure; that this is not easy is well known, for instance, from all those students in differential equations courses who lack an abstract concept of a function as a mathematical object and therefore fail to understand that a function rather than a number constitutes the solution to a differential equation.

There is a inherent difficulty in abstracting: How can we generate mental structures, which are so often linked to visual images, if they should represent relationships that are removed from the concrete objects which they were originally linked to? What is the role of visualization in the process of abstraction? Again, there is no definite answer to this question. Visual images are usually global and stress structural aspects. Therefore, if appropriate visual images can be found, they are likely to be of great help to students engaged in abstracting. The well known infinite row of domino stones as a model for mathematical induction is a case in point. It incorporates exactly those features that are common to all inductive processes and it does this in a manner that exhibits the structure of these common features. For instance, if one stone falls, then so does the next one; this is so at any place in the whole row of stones; therefore, if one falls, not only the next one but all following ones fall, each one causing the next one to fall. Furthermore, and this is central, it is obvious that the if-then-statement does not make the stones fall, but that in addition to the if-then-statement being correct, one, not necessarily the first, stone needs to fall. The picture of the domino stones contains the relevant elements of induction without many extraneous features. It captures the structure of the entire process, globally, into a single entity. Such a visual image undoubtedly helps students in building and strengthening their mental representation of induction. It happens, however, that visual models appropriate for abstract mathematical concepts are non-existent, incomplete or even misleading, and then care must be exercised. A detailed investigation of visually supported abstraction has been reported by Kautschitsch (1988); he found that dynamic visual sequences were strongly supporting abstraction because of their analogy to sequences of actions.

4. RELATIONSHIPS BETWEEN REPRESENTING AND ABSTRACTING (IN LEARNING PROCESSES)

Frequently, the concrete properties we would like to abstract from are linked to particular representations of an abstract mathematical concept; functions are a case in point, but so are vectors, vector spaces, groups, fields, C*-algebras, categories and functors. The properties and relationships of the abstract concept are the representation-independent ones.

Representing and abstracting are thus complementary processes in opposite directions: On the one hand, a concept is often abstracted from several of its representations, on the other hand representations are always representations of some more abstract concept. When a single representation of a concept is used, attention may be focussed on this instead of the abstract object. However, when several representations are being considered in parallel, the relation to the underlying abstract concept becomes important. Often representations are needed to carry out some specific work with the concept; for instance, group representations rather than abstract groups are used to carry out group theoretical calculations. This need for concrete representations in order to carry out some specific thought process is not purely mathematical. There is a parallel cognitive need: The thinking of many mathematicians and mathematics students is enhanced if they are able to place themselves mentally in a particular representation, e.g. a visual one. It is even more enhanced, when they are able to use several representations in parallel. Again, there is a complementarity, this time between the mathematical and the cognitive aspects of representing mathematical structures.

Both these complementarities, the one between abstracting and representing and the one between mathematical and mental representations, may be and have been put to didactic use in learning processes. Learning processes may then be seen as consisting of four stages:

- Using a single representation,
- Using more than one representation in parallel,
- Making links between parallel representations,
- Integrating representations and flexible switching between them.

In stage one, processes start from a concrete case, a single representation. But in learning the function concept, for example, students usually meet several representations (graphical, tabular, algebraic, arrow diagrams, ...). In the second stage, thus, several representations of the same mathematical object are used in parallel. The difficult process of transition to the abstract concept depends in an essential manner on the links between the representations that are formed. The establishment of these links constitutes the third stage. Strong links allow students to switch representations, which in turn makes them aware of the underlying concept and is thus likely to positively influence abstraction. At the fourth stage, a process of integration between the different representations is happening, a synthesis of the kind that has been shown above to be partial to the process of abstraction: the links, the relationships, the common properties remain to form the abstract concept, whereas the representation specific aspects retract to the background. Once this process has been completed, one has formed an abstract notion of a given concept, one somehow "owns" that concept. When one then needs to solve a problem in which this concept occurs, one will often need to go back to one (or several) of its concrete representations. The wonderful thing about the abstract concept is that one is able to do precisely that, and to do it in a controlled manner: One has control over the representations one wants to use.

The use of several representations to help students make the transition from a limited concrete understanding of a certain topic to a more abstract and flexible understanding, has been investigated by Kaput and his co-workers (1988). While dealing with ratio and proportion, they used concrete, visual, computer-implemented representations whose design was built on a cognitive basis. They called their approach a "concrete-to-abstract software ramp". The functions curriculum described in section 2.2 is another example: It was systematically built towards the establishment of links between different function representations and, as results from cognitive research have shown, this eventually lead most of the students in the experiment to an abstract understanding of the function concept.

For some particularly gifted students, such carefully designed constructions of concepts appear to be superfluous; the eight year old Terence Tao (Clements, 1983) is one such case: in spite of his age, he was able to learn directly about abstract algebraic structures, and concrete representations tended to disturb him, if anything. When asked whether a given structure was a ring, he immediately realized that he only needed to prove three things and proceeded to prove these with maximal reliance on earlier proved structural results. A similar situation may pertain to advanced mathematics students who have had the opportunity to acquire considerable experience with abstraction; this experience is likely to make some of the above stages superfluous and, for complex mathematical structures, even a hindrance to abstraction; as has been pointed out above, abstraction from one, or even from zero cases, may be easier for such students. But most students taking college and high school mathematics courses do not belong to this category, and for them abstracting is probably the most demanding of all advanced mathematical processes.

5. A WIDER VISTA OF ADVANCED MATHEMATICAL PROCESSES

The processes of representing and abstracting which have been discussed in some detail are among the most important ones for advanced mathematical thinking; nevertheless, they are only some among many processes which should and do occur as interacting links in chains that may also include discovering, intuiting, checking, proving, defining and others.

Discovering or rather rediscovering relationships, for instance, is often considered among the most effective ways for children to learn mathematics. To some extent, this effectiveness may be attributed to the psychological aspects of the process of discovery: the personal involvement, the intensity of the attention, the feeling of achievement and success. Learning by discovery, however, is time-consuming, and this is one reason why teachers, especially teachers of more advanced mathematics, tend not to use it. But the central question is whether learning by discovery in the early and middle stages of mathematical education develops reasoning processes which make later learning so much more efficient that there is, in the long run, no time loss. (Here, efficiency must be measured not only in terms of topics covered but also in terms of depth of understanding.) More generally, to what extent could more stress on processes and less on content enable our students to learn abstract mathematics in speedy, independent, and understanding ways?

Intuiting, i.e., apprehending by intuition, by immediate direct cognition without evidence of rational thought, has a central role in any sequence of processes that starts from discovery; the role of intuition will be discussed further in chapter 12 by Tirosh. Let us here only mention its close links to processes of visualizing. For instance, being interested in the intrinsic description of curves in terms of curvature (g) and arc-length (s), I was recently looking at ellipses; on the basis of the periodic increase and decrease of the curvature when proceeding around the ellipse, my visual intuition told me that a good try could be a trigonometric function of the form $g=A+B\cos(ks)$ with suitable constants A, B, and k. By the way, this intuition was soon proven wrong by more detailed analysis that included checking particular cases.

Checking means taking actions to convince oneself that a result indeed does answer the question that was asked, and does answer it correctly. One useful way of checking is to use an inverse procedure, such as differentiating to check whether one has correctly found a primitive function. All too often, checking is not seen by students as an essential part of mathematical activity. Although checking could give them a lot of security, most students appear not to be very interested in this security. This could and should be changed by transferring more of the responsibility for learning processes from the teacher to the student, in line with the independence mentioned in connection with discovering. Giving students open-ended activities rather than one-minute exercises to work on, is one step in this direction.

Discovering, intuiting and checking, however, are only the beginning of a sequence of mathematical processes – the goal remains understanding of abstract relationships. Students' activity must therefore proceed from here to the more formal processes of defining and proving, which will be analyzed respectively in chapter 5 by Vinner and chapter 13 by Alibert & Thomas.

It will be seen in the following chapters that many features of these processes need to be made very explicit to students, down into the smallest details. This does not mean that students should be told about these details, but that student activities have to be designed with these details in mind, in such a way that students realize them. These may, but need not, be details of mathematical facts or relationships; more often, they are parts of the processes. For example, with respect to switching representations, students must be made conscious of their act of pulling information out of one function representation and using this same information in another one (see Schwarz, Dreyfus & Bruckheimer, 1990, for details on how this can be achieved). Similarly, students need to become conscious of the interactions between processes such as representing and abstracting. The working mathematician is using many processes in short succession, if not simultaneously, and lets them interact in efficient ways. Our goal should be to bring our students' mathematical thinking as close as possible to that of a working mathematician's. Understanding advanced mathematical processes and their interplay is a necessary prerequisite for achieving this goal.

ACKNOWLEDGEMENT

This chapter has profited from numerous discussions with friends and colleagues, among whom were Michèle Artigue, Bernard Cornu, Ed Dubinsky, Janet Duffin, Jim Kaput, Uri Leron, John Mason, Baruch Schwarz, Pat Thompson, and Shlomo Vinner. But more than any others, Ted Eisenberg, David Tall, and Dina Tirosh have helped shape the chapter with well-taken comments and suggestions based on their careful reading of earlier versions.

MATHEMATICAL CREATIVITY

GONTRAN ERVYNCK

Creativity plays a vital role in the full cycle of advanced mathematical thinking. It contributes in the first stages of development of a mathematical theory when possible conjectures are framed as a result of the individual's experience of the mathematical context; it is also plays a part in the formulation of the final edifice of mathematics as a deductive system with clearly defined axioms and formally constructed proofs. It is an essential factor in research mathematics when new ideas are formulated in a manner previously unknown to the mathematical community. Yet it is *external* to the theory of mathematics. It is a human activity, a meta-process, which acts upon and generates new mathematics. As such it is often viewed as a mysterious phenomenon. Most mathematicians seem to be not interested in the analysis of their own thinking procedures and do not describe how they work or conceive their theories. Only a few (such as Poincaré, Hadamard) explicitly attempt to describe ideas related to mathematical creativity. The best known reference (at least to mathematicians) is probably Hadamard (1945), which has been followed recently by Muir (1988).

The present chapter will not attempt to give an exhaustive description of the nature of mathematical creativity and how it works. From a somewhat closer look at the aspects of different kinds of mathematical activity as an heuristic procedure to register examples of mathematical creativity, we derive some striking characteristics of the phenomenon and frame a tentative definition. The reader is invited to activate his/her own imagination and attentiveness and to rectify deficiencies in the text with their own personal observations.

1. THE STAGES OF DEVELOPMENT OF MATHEMATICAL CREATIVITY

Mathematical creativity does not, presumably cannot, occur in a vacuum. It needs a context in which the individual is prepared by previous experiences for the significant step forward in a new direction. Such preparation occurs through previous activities which form an appropriate environment for creative development. We hypothesize that the context for creativity is set by a preparatory stage in which mathematical procedures become interiorized through action before they can be the objects of mathematical thought. Preliminary to this may be an initial stage where the procedures might be used without even a full appreciation as to their theoretical status.

Stage 0: A preliminary technical stage
We hypothesize that genuine mathematical activity may be preceded by a preliminary stage consisting of some kind of technical or practical application of mathematical rules and procedures, without the user having any awareness of their theoretical foundation. We refer here to the art of the craftsman who applies a set of mathematical procedures as a toolkit

providing him with the necessary tools to manufacture his product. The justification for these procedures is that it has been checked empirically that they work, in the sense that a correctly applied rule always yields the desired result. An example of such a practical procedure is the rule used in Ancient Mesopotamia and Egypt to stake out a right angle: they used a rope with marks dividing it into three parts of length 3, 4 and 5. Forming the contour of a triangle, they obtained a right angle between the sides of length 3 and 4. This preparatory stage has become part of modern theories of mathematics learning, for instance, the "tool-object" dialectic of Douady (1986) which proposes that an idea should be introduced first as a tool as part of a problem-solving activity, to become part of the individual's experiential cognitive structure before being reflected on as an object in its own right.

Stage 1: Algorithmic activity

At this stage procedures are used to carry out mathematical operations, to calculate, manipulate, solve. Algorithmic activity is essentially concerned with performing mathematical techniques. Examples of such techniques are: application of an algorithm, working out formulae, factorizing a polynomial, calculating an integral, computational activities involving computer programs such as in numerical methods for solving differential equations. A characteristic of such activities pertaining to this first stage is that they need to be quite explicit. All intermediate steps have to be considered, at least implicitly; if not, a serious error may occur and totally invalidate the result. In the case of a computer algorithm, no steps, not even trivial ones, may be forgotten. There is no regeneration of missing steps in an algorithm.

Such activities are an acceptable part of advanced mathematics because they may be seen as part of an overall theory, created in accordance with the principles of the higher activities to be described in stage 2. We hypothesize that algorithmic activity is an essential part of the *learning* of mathematics because such processes must be interiorized and become routinized before they can be reflected upon as manipulable mental objects in a higher order theory. (See chapter 7 for a discussion of reflective abstraction, in which a process is encapsulated as a concept). As with the tool-object dialectic, it is essential that the tool become familiar *in action* before it becomes the focus of reflective activity.

Stage 2: The creative (conceptual, constructive) activity

It is at this second stage that true mathematical creativity is likely to occur and act as a motive power in the development of a mathematical theory. A non-algorithmic decision is taken in a manner which seems to signify a bifurcation of the underlying concept structure. Mathematical creativity is the ability to perform such steps. The decisions that have to be taken may be of a widely divergent nature and always involve a choice, such as a choice of a certain concept to be defined (for instance, as in Hausdorff's choice of the notion of an *open* set, which proved to be of tremendous importance in major parts of mathematics) or the decision to state and prove a theorem. The latter requires two distinct creative steps: the choice of adequate hypotheses such that the resulting conclusion is of value within a wider theory, and the actual deduction from the hypotheses to establish the proof of the theorem. Note that the initial choice demands also a formulation of the possibilities from which the choice is made. It is in such complex activities that mathematical creativity comes to the fore.

In order that mathematical creativity should be activated, there is no need to have a formal theory at one's disposal; the most active part of creativity acts at the intuitive level in a spirit of regeneration and renovation. Davis & Hersh (1981) suggest that it comes through a passage from the coarse (the intuitive) to the fine (the formalized).

What is essential in the individual is a state of mind prepared for mental activity that relates previously unrelated concepts. It is often observed to occur after a period of intense activity involving a heightened state of consciousness of the context and all the constituent parts. And yet it is more likely to bear fruit when the mind is subsequently relaxed and able, subconsciously, to relate the ideas in a manner which benefits from quiet, unforced, contemplation.

High levels of creativity demand a subtlety of mental structure that is tuned to be able to resonate with underlying patterns that may fail to come to light in less refined circumstances. We illustrate this by looking at an example of a problem which can be solved at different levels of mathematical sophistication. The three levels may be characterized in a very typical manner, based on the status of the method used in the solution. This classification proves to be parallel to the description of the stages of development of creativity just described. A first (low) level relies heavily on the application of an algorithm; the creativity involved requires only recognition of the overall positioning of the problem in the whole of mathematics and the construction of the appropriate model (for instance, a system of linear equations or a truth table). A second (higher) level abandons the straightforward application of the algorithm, but is based on direct reasoning inside the mathematical model. Some insight and intuition are needed to develop the right method of solution. The environment (the model) is still borrowed from a general theory, but solving the problem at hand is done by inferring directly from the given situation. A third level (the highest) abandons the model completely, reasoning outside a formalized theory, constructing a solution *ab ovo* by an intelligent inspection of what is stated in the problem.

- *Problem*: A man was a child for one sixth, a young man for one twelfth and a bachelor for one seventh of his life. His son was born five years after his marriage and died four years earlier than his father. The lifetime of the son was half of the lifetime of the father. How old was the father when he died?

- *Solution at a low level of mathematical creativity*

  It is enough to realize that the problem is subject to strict conditions and can be modelled in a system of algebraic equations. The abilities involved are the introduction of the necessary unknowns and the formulation of the equations. Let

  x = the age of the father at his death;
  y = the time the father has been married;
  z = the age of the son at his death.

  A careful translation of the problem gives the equations

  $$\frac{x}{6} + \frac{x}{12} + \frac{x}{7} + y = z$$

$$5 + z + 4 = y$$

$$z = \frac{x}{2}$$

and a knowledge of the solution of linear equations gives the solution $x=84$.

- *Solution at a higher level of mathematical creativity*

An endeavour to formulate a concept image of the internal structure of the problem may yield the awareness that the situation has a linear model. As the lives of human beings occur in time, a time axis is all that is required to represent all the events occurring in the problem. Moreover, all events refer to precise moments which may easily be located on the time axis. A simple graphical representation:

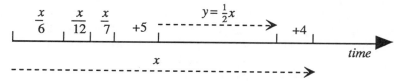

Figure 1 : Visualizing the solution

gives (suitably interpreted) the following equation:

$$\frac{x}{2} = (\frac{1}{6} + \frac{1}{12} + \frac{1}{7})x + 9$$

and the solution $x=84$.

- *A third level solution*

A much more sophisticated method for solving the problem is based on intuition, experience and some plausible (in the sense of Polya) assumptions embedded in the nature of the problem.

• A first hypothesis is the assumption that the age is generally expressed as a positive integer in the range 0 to 100. A plausible starting point is to look for an integer solution.

• A second step is the assumption that the fractions $\frac{1}{6}$, $\frac{1}{12}$, $\frac{1}{7}$ occurring refer to periods in the life of the father which are probably whole numbers as well.

• A decisive step is to realize that the denominators, 6, 12 and 7 have few common multiples between 0 and 100, hence it may be valuable to compute the lowest common multiple – indeed the only common multiple in this range – which is 84. Verification confirms that 84 is the (only) solution.

This higher level of mathematical activity involves a highly tuned experience with number theory as well as an insight into the working methods of the problem-poser. It illustrates the unusual pathways that might be taken by the creative mathematician in solving an old problem in a new way. But to be able to analyse the role of creativity of new mathematics in new contexts we must first consider the nature of advanced mathematics as a major goal of mathematical creativity.

2. THE STRUCTURE OF A MATHEMATICAL THEORY

It is essential to have an overall view of the structure of mathematics as a mental construct before concentrating on the nature of the creative processes that bring it into existence. We see a formal theory of mathematics as a framework consisting of definitions of concepts and relations between the defined concepts, the latter being of a very particular kind: the relations emerge from the implementation of very strictly prescribed (deductive) rules. This entails the necessity to determine (to define) the concepts is a very precise manner. The concepts may be thought of as the nodes of a network and the relations are directed arrows connecting the nodes. Moreover, the network has an additional feature: the connections are ordered proceeding from the logically basic nodes towards the more complicated ones. Mathematical creativity involves both the vision to build up parts of such a framework by conjecture and argument and also to refine the structure into a mathematically deductive framework.

We suggest that an act of creativity requires the realization of at least one of the following objectives:

(i) to create a useful new concept, where 'useful' means favourable to the further unfolding of the theory at hand;

(ii) to discover a formerly unnoticed relation between two (or more) nodes, with the required ordering;

(iii) to construct a useful ordering: to organize a part of a theory such that its logical, deductive order becomes more apparent.

The specification of a successful set of axioms for a previously unaxiomatized theory (as with group theory) may be considered as an instance of mathematical creativity where all three objectives have been realized.

3. A TENTATIVE DEFINITION OF MATHEMATICAL CREATIVITY

Examples of creativity in mathematics are: the ability to formulate a valuable definition using concepts which assure the usefulness of the defined object in the subsequent theory; or the formalization of a basic idea borrowed from the physical context which was initially at the base of the mathematical problem. Hence, we look at mathematical creativity essentially as the ability to create mathematical objects, together with the discovery of their mutual relationships. This activity is considered here as different from, and even opposed to, algorithmic mathematical objects (the "first stage" mentioned earlier).

A tentative definition might be:

> Mathematical creativity is the ability to solve problems and/or to develop thinking in structures, taking account of the peculiar logico-deductive nature of the discipline, and of the fitness of the generated concepts to integrate into the core of what is important in mathematics.

4. THE INGREDIENTS OF MATHEMATICAL CREATIVITY

The working procedures of mathematical creativity are intimately linked with the stages discussed in section 2. They are essentially a working-out of the impulses which steer the creativity of the working mathematician and operate generally in the following order:

(1) study, yielding familiarity with the subject,

(2) intuition of the deep structure of the subject,

(3) imagination and inspiration,

(4) results, framed within a deductive (formal) structure.

It is the effort involved in studying and becoming familiar with the subject that sets in the mind conceptual structures that contain the potential for mathematical creativity. Intuition is the product of the action of these conceptual structures on newly acquired data. As we saw in chapter 1, intuition can be honed and polished into a refined tool. The more refined the mental structure, the more likely it is to produce refined intuitions. It is by reflection on the deep structure of the subject that such intuitions may lead to the imagination and inspiration which formulate the required results, at first in an imperfect form, but then honed by reflection into formal deductive order.

5. THE MOTIVE POWER OF MATHEMATICAL CREATIVITY

The power of mathematical creativity results from the interaction of a certain number of elements which may be listed as follows (although there is no reason to believe that the list is exhaustive):

- *Understanding*: the ability to regenerate the steps of the mathematical creativity of the author(s) of a theorem, a part of a theory ... Mathematical creativity is based on, and brings with it, a simultaneous deepening of understanding and insight in a concept.

- *Intuition*: the formation of concept images which are sufficiently close to the formal concept to allow the conception of plausible conjectures. Intuition enables the mathematician to perform a fruitful selection as well (see below). Other factors of equal importance, related to intuition and acting as driving forces in the process of mathematical creation are imagination, mathematical phantasy and curiosity (Dieudonné, 1974).

- *Insight*: the driving force required to move towards a formulation of new knowledge. This involves a refocussing of interest and a reorientation to consolidate what is important, and even more, to envision will be important in the future.

- *Generalization*: the ability to generalize is linked with insight because it depends heavily on the ability to foresee what will be important in the future. Generalization is a form of mathematical creativity, but sometimes only a weak form: generalizing a theory is sometimes hard, sometimes straightforward. Sometimes it may be an illusion: as any finite group has a representation as a group of permutations, the generalization of the theory of permutation groups of Galois and Jordan in the theory of finite groups is only a reformulation, though undoubtedly a more handsome embodiment, of the former.

We see these four ingredients being parallel to the four topics mentioned in the previous section. By understanding, we mean not just the *instrumental* understanding involved in being able to carry out processes, but the *relational* understanding, in the sense of Skemp (1976), which involves a meaningful grasp of the relationships between the concepts. Even this is not enough, for it suggests a meaningful relationship between the concepts in the context which they are currently known. Creativity demands an extension of this context in a way that has not before been conceived. It therefore requires the individual to create new ideas and to put old ideas together in a new way. It is not something which can be carried out on demand.

"The philosopher's stone can only be found when the search lies heavily on the searcher. Thou seekest hard and findest it not. Seek not and thou willst find."

It requires relaxation and incubation in the sense of Poincaré (see page 15). Given good preparation and good fortune, the incubation may provoke intuitions that lead to the fundamental insights that break through to give the creative leap.

The latter may be a generalization of previous knowledge, which means the extension of current schemas to a broader context. In chapter 1 we saw that there were two fundamentally different kinds of generalization: the expansive generalization, which broadens the applicability of the theory without changing the nature of the cognitive structure, and the reconstructive generalization which requires the knowledge structure to be reorganized. Whilst the former may be relatively easy, even when it occurs creatively for the first time, the latter involves a cognitive transition of great difficulty which requires special personal qualities of character to succeed in the struggle.

6. THE CHARACTERISTICS OF MATHEMATICAL CREATIVITY

In making the great leap of mathematical creation, we see certain characteristics coming to the fore. Mathematical creativity is:

- *Relational* (in the sense of Skemp). It stimulates through interaction: it establishes a conceptual link between two or more concepts, such that a new idea emerges which integrates different aspects form the initial concepts into a single one. Interaction of ideas in the mind of the mathematician is perhaps the most important driving force of mathematical creativity. Mathematical ideas and concepts arise as mobile building blocks and combine (if the subject is not mathematically totally insignificant) to form some new configuration. If the configuration is favourable, it enters into the theory. This has already been described by Poincaré.

A deeper view of the process entails the question:

is mathematical creativity acting just as mutations in biology?

A mathematical mutation occurs when a chain of ideas undergoes a restructuring, maybe in one single place. Among all restructurings some are useless, others are useful. Some survive, others are eliminated although they are entirely correct from a formal viewpoint. An example of such a case is the theory of cubics, algebraic curves of degree three, developed as a generalization of the theory of conics; this theory was developed in the nineteenth century, but is seldom taught today.
We therefore also see that mathematical creativity is:

- *Selective.* This analogy with biology arises through the struggle for life amongst mathematical concepts, with a natural selection and survival of the fittest. For example, the several theories of integration established at the end of the nineteenth and the beginning of the twentieth century to generalize the Riemann integral entered into competition with each other and finally the Lesbegue integral survived to dominate mathematical analysis (Van Dalen & Monna, 1972).

Selectivity gives rise to a related criterion:

- *Fitness.* This is a qualifying criterion for the value of definitions and theorems and sets of axioms in mathematics. The well-known estimation by Stanislas Ulam of the 200,000 yearly produced theorems makes it clear that a sieve seems very necessary. In fact, the sieve exists, and in the first place does not consist of the referees of the numerous journals, but acts spontaneously and unconsciously in time, through the action of the struggle for mathematical life and the survival of the fittest ideas.

Finally, mathematical creativity must lead to new ways of handling the complexity of the relationships between more complex concepts. It does this by encapsulating new structures into single objects which are easier to manipulate mentally. It is therefore:

- *Condensing*. Mathematical creativity includes the ability to choose the appropriate wording and symbols for the representation of mathematical concepts. The importance of symbolic representations in mathematics cannot be overestimated. Well-chosen symbols allow for a condensation of several aspects of one concept into a single whole which is evoked every time the symbol occurs in a text. In this manner the use of the symbol frees "memory space" in the mind which becomes available for other, till then unknown or unclear concepts.

7. THE RESULTS OF MATHEMATICAL CREATIVITY

After the process of mathematical creativity, there are various qualities that the new ideas must exhibit in order that they might be accepted and survive in the mathematical community at large. MacLane (1986) suggests a number of criteria which are required so that the new idea can be labelled "good mathematics". It must be:

- *Illuminating*. This seems to be a necessary characteristic of mathematical creativity. Good mathematics should be of help in understanding. A result that obscures is not creative, or is creativity used in an inappropriate direction, for example through indulging in long technical calculations. For the same reason we say that mathematical creativity in the first stage (algorithmic activity) is very low.

- *Deep*. Mathematical creativity is supposed to uncover hidden relationships. A deep result is not necessarily difficult to prove, but it is usually wide in its relevance and application.

- *Responsive or fruitful*. The successful product of creativity is based on former results and often responds to current needs. If it is to survive, it also provides a basis for future development so that it remains an essential part of living mathematics.

- *Original*. There should be something unexpected in the results, something new in the field, if it is just a rearrangement of known results, there will be strong doubts concerning the creative aspect of the achievement.

In addition, there are subtle qualities of surprise, even humour, which cause a mathematical result to appeal to a professional mathematician. The following example illustrates the latter (though it is not put forward as an example of particularly deep mathematics). It is an inference which occurs using well accepted methodology (use of axioms, logical deductions, and so on) and is analogous to the usual reasoning in mathematical papers, but the result of the inference is strange and unexpected.

The problem (from Wille, 1984) runs as follows:

In the teaching of geometry, how shall the teacher proceed in order to draw a "general" triangle on the blackboard?

The problem is ill-posed as long as we have no agreement on what a general triangle is; hence a clear description of the term "general" is required. Thus the first step in mathematical theory is encountered: the formulation of clear definitions. As we want to serve the purposes of teaching for mathematically inexperienced students, it is quite acceptable to say that "general" means a triangle without any particular geometrical property. Moreover, granted the students lack of mathematical experience, we claim that it must be seen with the naked eye that the triangle is general. For example, a triangle with angles 89°, 45°, 46° will be perceived by the students as a right-angled, equilateral triangle, and so does not fit our requirements. The didactical principle underlying these concerns is that mathematical concepts are understood and developed on the base of concept images which are present in the learner's mind. A correct concept image may be generated by a selection of appropriate examples; hence, in the case of geometry, the selection of the pictures to be drawn on the blackboard may yield a correct understanding (or not) of the formal ideas.

The claim that the students recognize the generality of the triangle by naked eye requires empirical investigations. Based on experiments, a model has been constructed that describes the ability of the human eye to distinguish between angles of different sizes. It appears that inability to recognize a second angle as different from a first given one is normally distributed according to the difference between the two angles. Experimentally the standard deviation is $\sigma = 5.77°$, with a 99% certainty obtained by a difference interval of size $2.6\sigma = 15°$. We therefore adopt the statement that a triangle is "general" if it is considered as such by 99% of the students.

This leads to the following axiom system which formalizes the conditions established in the previous paragraphs:

Axiom I: The triangles not isosceles.

Axiom II: The triangle is not right-angled.

Axiom III: Two angles differing by less than 15° are considered as being equal.

With these axioms we may prove the following remarkable theorem:

Theorem There is, up to similarity (axiom III) precisely *one* general triangle with acute angles, namely, 45°, 60°, 75°.

(We note that there are an infinite number with one obtuse angle.)

Proof Let the triangle have angles A, B, C where $90 > A > B > C$.
As A differs from 90° by at least 15°, we have $A = 90 - 15 - a = 75 - a$ where $a \geq 0$.
Similarly, B differs from A by at least 15°, so

$$A = 75 - a - 15 - b = 60 - a - b$$

where $b \geq 0$ and finally

$$C = 60 - a - b - 15 - c = 45 - a - b - c \text{ where } c \geq 0.$$

But the angles add up to 180°, so

$$180 = 75 - a + 60 - a - b + 45 - a - b - c = 180 - 3a - 2b - c$$

and the non-negative numbers a, b, c satisfy

$$3a + 2b + c = 0,$$

hence $a = 0, b = 0, c = 0$ and

$$A = 75, B = 60, C = 45,$$

as stated.

8. THE FALLIBILITY OF MATHEMATICAL CREATIVITY

A major characteristic of mathematical creativity which distinguishes it from the generally accepted qualities of a mathematical theory is that it is sometimes fallible. It puts together new ideas in a way which may prove to be insightful, or may equally lead to error. There is no guarantee that theorems may be formulated correctly, or even that such theorems are accompanied by correct proofs. Famous examples are early proofs of the Four Colour Theorem, the numerous "proofs" of the fifth postulate of Euclid and the recent "proof" of the Poincaré Conjecture which seemed plausible for some months before a flaw was discovered.

Given the view of Lakatos (1976), mathematics does not proceed in a Vauban-like manner, making step-by-step sure advances in a pre-determined direction, but like the daring exploits of the cavalryman of the advance guard, the forays into new territory may be flawed. Mathematical thinking, as opposed to the reflected organization of mathematical thought, is a creative activity that brings with it the possibility of human error. Indeed the very possibility of error is what makes the major advances such monuments of human success.

9. CONSEQUENCES IN TEACHING
ADVANCED MATHEMATICAL THINKING

The fallibility of this vital stage in mathematical thinking is something that students may find hard to accept. Their whole mathematical training is usually accompanied by the provision of algorithms that provide certainty to solve a given class of problems, and with it the (false) belief that, given sufficient time and study, there will be an algorithm that will solve any given problem. When they study differential equations, they see the solution of various types of equation: separable equations of first degree, those that can be solved using an integrating factor, or a power series approach, the special case of simple harmonic motion, then higher order differential equations with constant coefficients. It will come as a surprise to such students that the subset of differential equations that can be solved is, in a genuine cardinal number sense, an insignificant minority of all differential equations.

Students are so often given the impression that, in mathematics, all is logical, certain, accurate, provable, amenable to clear explanation. Yet mathematical creativity is none of these things. It offers a major difference between the actual working practices of research mathematicians and the facets of the mathematician's art that are selected to teach to the next generation.

We have seen that there are certain requirements of mathematical creativity which seem to preclude its operation in all but the most gifted. In particular it requires a sophisticated understanding of mathematics in a given context to make creative developments which extend known theory. Clearly we should not expect students to (re-)invent what has taken centuries of corporate mathematical activity to achieve. Yet if we do not encourage them to participate in the generation of mathematical ideas as well as their routine reproduction, we cannot begin to show them the full range of advanced mathematical thinking.

Such approaches are already beginning in elementary schools where children are being asked to carry out extended mathematical investigations starting from a context which is, to them, novel. For them, such an enterprise is creative. It provides an activity which is complementary to traditional methods of learning mathematics without in any way replacing them. It allows children to begin to explore, to conjecture and test, to formulate and prove, in ways which give deeper meaning to mathematical processes.

In this way, at a time when the content and approach to elementary mathematics is becoming more clearly prescribed in national curricula and national standards, there are complementary moves to encourage younger children to play their own part in knowledge generation, to make conjectures, to expect errors, to need to check, to convince, to prove. In a society which is fast changing, such flexible thinking beyond the mere application of algorithms is becoming not just desirable, but increasingly necessary. Creativity at only the lowest level is no longer acceptable.

In the next chapter we will see that the apparently unimpeachable bastion of mathematical truth, the formal proof, is in practice context bound and dependent upon stylistic conventions of the mathematical community. It is therefore somewhat more fallible than mathematicians may care to admit. The wider appreciation of the full range of advanced mathematical thinking, including knowledge generation and creative problem-solving through conjecture, debate and proof is therefore an objective which is worthy of consideration. In chapter 13 we will return to the question of conjecture and debate in the creation of mathematical proof. Within this broader framework of advanced mathematical thinking, we therefore see mathematical creativity, so totally neglected in current undergraduate mathematics courses, as a worthy focus of more attention in the teaching of advanced mathematics in the future.

CHAPTER 4

MATHEMATICAL PROOF

GILA HANNA

The hallmark of the mathematics curriculum adopted in the sixties was an emphasis on formal proof. Among the manifestations of this emphasis were an axiomatic presentation of elementary algebra and increased attention to the precise formulation of mathematical notions and to the structure of a deductive system.

Indeed mathematics itself had grown tremendously since the beginning of the twentieth century. Entire new fields had come into being in the first half of the century: modern mathematical statistics, the theory of games, queuing theory, graph theory, and techniques such as linear programming, often included in the general category of operational research, which had gained prominence through their successful application during World War II.

The growth of mathematics was accompanied by change of outlook on the part of practising mathematicians. The work of the logicist, formalist, and intuitionist schools on the foundations of mathematics had given an impetus to the concern for precision in definition and for the careful use of language. Also, the axiomatic approach, which these schools shared despite their many differences, had become a common denominator of most mathematical endeavours. The new status of deductive rigor as a standard in mathematical work was stated clearly by the prominent French mathematician Dieudonné (1971):

Hence the absolute necessity from now on for every mathematician concerned with intellectual probity to present his reasonings in *axiomatic* form, i.e., in a form where propositions are limited by *virtue of rules of logic only,* all intuitive "evidence" which may suggest expressions to the mind being deliberately disregarded. (p. 253)

Many important mathematicians saw the axiomatic method not only as the prescribed form for each individual mathematical discipline, but also as a means of consolidating many previously unconnected disciplines into a small number of "mathematical structures". This point of view was promulgated by the group of influential French mathematicians who wrote under the name of Bourbaki. The Bourbaki group exerted a great deal of influence internationally on mathematical research. The focus of the group, apart from its attention to newer mathematical subject matter, was on what has been called the "Bourbaki approach": a formal, abstract, and rigorous approach, emphasizing precise definitions and formal proof.

This chapter discusses the origins of this emphasis on formal proof and considers its limitations as a focus for advanced mathematical thinking in light of those aspects of mathematical practice which complement and go beyond formal proof.

1. ORIGINS OF THE EMPHASIS ON FORMAL PROOF

The curriculum revolution of the sixties was predicated upon a number of beliefs, one of which was that formal proof is the most important characteristic of modern mathematics. This view was no doubt due in large part to the impressive work done during the first half of the century in clarifying the very foundations of mathematics, work which had demonstrated the enormous power of formal systems constructed step by step from a base of definitions, axioms and rules of inference.

The brilliant mathematicians who were so united in a desire to lay a firm foundation for mathematics were by no means united in their approach to this task. But though the schools of thought which came to be known as logicism, formalism and intuitionism differed greatly in their philosophical accounts of mathematics and even in their criteria for the validity of a proof, they did share an emphasis on the importance of formal proof, and it is this emphasis, rather than the differences among the schools, that has so greatly influenced the mathematics curriculum.

The central assertion of logicism is that mathematics is part of logic. Accordingly, the aim of the logicists was to produce the corpus of mathematics without introducing concepts indefinable in logical terms or theorems which cannot be proved from the primitive sentences of a logical calculus using its tightly-defined rules of proof. Thus formal proof played a central role in the logicist agenda.

This was true of the formalist effort as well. In fact, the thesis of the formalist school was precisely that mathematics is a science of formal systems: that it deals with the manipulation of strings of symbols to which no meaning need be assigned. In the formalist view, the validity of any mathematical proposition rests upon the ability to demonstrate its truth through rigorous proof within an appropriate formal system.

The intuitionists, too, assigned importance to formal proof. Their differences with the logicist and formalist schools centred upon the types of proof which should be admitted as valid, with the intuitionists taking a more restrictive view. Intuitionism is the belief that mathematics and mathematical language are two separate entities, mathematics being essentially a languageless activity of the mind. Mathematical activity then consists of "introspective constructions", rather than axioms and theorems. But for the intuitionist the assertion of a mathematical proposition was equivalent to the assertion that there is a construction of a finite nature which produces the proposition – and such a construction had to obey rules of rigour.

2. MORE RECENT VIEWS OF MATHEMATICS

In the last two decades several mathematicians and mathematics educators have challenged the tenet that the most significant aspect of mathematics is reasoning by deduction, culminating in formal proofs. In their view, there is much more to mathematics than formal systems. This view recognizes the realities of mathematical practice. Mathematicians admit that their proofs can have different degrees of formal validity – and still gain the same degree of acceptance. Mathematicians agree, furthermore, that when a proof is valid by virtue of its form only, without regard to its content, it is likely to add very little to an understanding of its subject and ironically may not even be very convincing.

In these more recent views, a proof is an argument needed to validate a statement, an argument that may assume several different forms as long as it is convincing. Proof has been described as a "debating forum" (Davis, 1986), as having "a certain openness and flexibility" (Tymoczko, 1986), and as possibly depending for its validity on "correct or reasonable social practice" (Kitcher, 1984).

In examining how proof in mathematics takes into account a social process and hence goes beyond the concept of formal proof often reflected in mathematics teaching, ideas advanced by Lakatos (1976), Kitcher (1984), Tymoczko (1986) and Davis (1986) will be discussed.

Lakatos has expressed the point of view that mathematics is fallible by its very nature. His account of mathematics is thus at odds with both logicism and formalism, undoubtedly influenced by their failures. Though mathematics is not an empirical science, Lakatos shows that its methods are very similar to those of the empirical sciences; he refers to mathematics as quasi-empirical. Mathematics, in fact, grows through an incessant "improvement of guesses by speculation and criticism, by the logic of proof and refutation" (Lakatos, 1976). Thus no proof is final, and indeed it is the essentially social process of negotiation of meaning, rather than the application of formal criteria from the outset, which leads to the improvement of a proof and its growing acceptance.

According to Kitcher (1984), to understand the development of mathematical knowledge one must focus on the development of mathematical practice: mathematical knowledge owes its growth to rational modifications to this practice. Mathematical practice has five components:

(1) a language,

(2) a set of accepted statements,

(3) a set of accepted questions,

(4) a set of accepted reasonings,

and

(5) a set of metamathematical views.

The latter component includes standards for proof and definition, as well as claims about the scope and structure of mathematics (p. 163). Thus in his view it is not only the corpus of mathematical results which develops, but also the very ways in which mathematics is done.

Further, Kitcher does not accept the a priorist view that mathematical knowledge is based on proof. He attacks the conception that "proposes to characterize the types that count as proofs in *structural* terms" (p. 36). It is furthermore historically incorrect to assume that change in mathematics has consisted only of the discovery of earlier mistakes and their replacement by new, correct demonstrations. In his view mathematical knowledge is always sensitive to peer challenges and is sustained, in part, by community approval of assumptions and techniques.

Citing examples from the work of Euler, Cauchy, Weierstrass and Newton, Kitcher concludes that mathematical proof is not always necessary to mathematical knowledge, and that it may not even be rational to attempt to accumulate a series of certainties; a demand for rigour may even be a hindrance to the growth of mathematics, because it impedes problem solving. Indeed, within the set of accepted reasonings (mentioned above as a component of mathematical practice), the most interesting are those which occupy an intermediate position in this set: the unrigorous ones.

Tymoczko is a philosopher of mathematics who thinks that what mathematicians actually do has a bearing on philosophical questions about mathematical knowledge. According to him, the concept of the "ideal mathematician", the totally rational agent who needs only follow formal deductive procedures to generate eternal and infallible knowledge, is not one which is helpful in the philosophy of mathematics (Tymoczko, 1986).

Tymoczko's account of mathematical knowledge centres upon the community. It views not only mathematics teaching, but also the concept of proof and the practice of proving theorems, as *public* activities. Regarding the concept of proof, he agrees with Lakatos that "proof ideas" are subject to criticism and even invite it. In his view

Mathematical proofs... generally have a certain openness and flexibility. They can be paraphrased, restated and filled out in various ways, and to this extent they transcend any particular formal system. We might say that an informal proof determines an open-ended class, or family, to use Wittgenstein's term, of more specific proofs.
(p. 49)

Tymoczko goes on to say that informal proofs are often convincing and can lead to new discoveries. They are codified in terms of simple proofs ideas and become the property of a network of mathematicians.

Davis (1986) states that a proof can play several different roles. It can serve as a validation, it can lead to new discoveries, it can be a focus for debate, and it can help eliminate errors. According to him, as to Tymoczko, the traditional philosophies of logicism, formalism and intuitionism are "private theories" that describe an ideal mathematics. But mathematics, being a social activity, requires a public theory.

In the real world of mathematicians, Davis believes, a proof is never complete and furthermore cannot be completed. Routine calculations will invariably be omitted. There will always be an appeal to intuition, to pictures. There will be some metamathematical objections, but such a part-proof will nevertheless be convincing because it is addressed to people who share a mathematical subculture in which an incomplete argument is understood, appreciated and seen as adequate. A typical college lecture in advanced mathematics will include formulations such as "it is easy to show", "you can verify that", "by an elementary computation which I leave to you", and so forth. It is considered perfectly proper to transmit mathematics in this elliptical way.

Davis is quite explicit in his view of formal proof:

There is a view of proof or a view of mathematics which I disagree with and which I think is a myth, which says that mathematics is potentially, totally formalizable, and therefore, one can say, in advance, what a proof is, how it should work, etc.
(p. 336).

3. FACTORS IN ACCEPTANCE OF A PROOF

Clearly the acceptance of a theorem by practising mathematicians is a social process which is more a function of understanding and significance than of rigorous proof. Indeed, the presence of any proof, rigorous or otherwise, is only one of several determining elements in acceptance. This process is by no means capricious; the community judges by certain criteria, as I will discuss. But the significance of a theorem for mathematics as a whole, and an understanding of its underlying concepts, play a much greater role in creating this acceptance than does the existence of a rigorous proof.

The development of mathematics and the comments of practising mathematicians suggest that most mathematicians accept a new theorem when some combination of the following factors is present:

- They understand the theorem, the concepts embodied in it, its logical antecedents, and its implications. There is nothing to suggest it is not true;

- The theorem is significant enough to have implications in one or more branches of mathematics (and is thus important and useful enough to warrant detailed study and analysis);

- The theorem is consistent with the body of accepted mathematical results;

- The author has an unimpeachable reputation as an expert in the subject matter of the theorem;

- There is a convincing mathematical argument for it (rigorous or otherwise), of a type they have encountered before.

If there is a rank order of criteria for admissibility, then these five criteria all rank higher than rigorous proof.

Perhaps the situation is best discussed in terms borrowed from Maslow's theory of social motivation (Maslow, 1970). Understanding, significance, compatibility, reputation, and convincing argument are "positive motivators" to acceptance: it is these factors which focus the attention of practising mathematicians on a new theorem and move them to its active acceptance, lifting it above the great body of equally valid but less attractive theorems which confront them in the mathematical literature.

On the other hand, the structural validity of the mathematical argument for a new theorem, that is, the actual or potential validity of its form as distinct from its content, is merely a "hygiene factor", a factor recognized as essential but taken for granted. There is a presumption that any convincing proof appearing in a reputable journal is in fact valid in terms of its form, or could be made so without violence to its content. The publication of a rigorous proof would provide no additional positive motivation for active acceptance, and in fact such a proof would not be examined at all in the absence of the motivating factors enumerated above.

4. THE SOCIAL PROCESS

The following discussion will establish the fact of a social process in acceptance, and the central role played in that process by the factors of understanding, significance, compatibility, reputation, and convincing argument. The Russian logician Manin is among those who have stressed the fact that the acceptance of a proof depends much more on a social process than on some ideal objective criterion:

> A proof becomes a proof after the social act of "accepting it as a proof". This is true of mathematics as it is of physics, linguistics, and biology. (Manin, 1977, p. 48)

Manin then goes on to explain that a new proof needs to be accepted and approved by other mathematicians – who often decide to refine and improve it. The scrutiny to which mathematicians subject a proof, he points out, is aimed more at weighing the plausibility of the results than at verifying the deductive process. It is only when they are skeptical of a result that mathematicians will put any great effort into discovering counter-examples. Manin cites this as the reason why the truth of a theorem in the eyes of the mathematical community becomes established indirectly, that is, not because the proof has been verified as error-free, but because the results are compatible with other accepted results and the arguments used in the proof are similar to ones used in other accepted proofs.

Of the estimated 200,000 theorems published yearly (Ulam, 1976), only a very few are actively accepted by the mathematical community. It is the theorems judged significant that have their proofs scrutinized, corrected, and refined, while the proofs of other theorems go unexamined. Clearly an alleged proof of Fermat's last theorem or the four-color theorem, when submitted by reputable mathematicians, would attract meticulous review, while the proof of a theorem of no apparent consequence is likely to be ignored, no matter how original or sophisticated the proof might be in its own right.

Indeed, as Davis (1972) notes, most proofs in research papers are never checked. Many of them are rife with errors, in fact. This is borne out by the many mistakes found in those published proofs which have been checked, and is also supported by the contention of a former editor of *Mathematical Reviews* that as many as half of the proofs published are false, though the theorems they purport to prove are essentially true. When an error is detected in the proof of a significant theorem, it is often the proof that is changed, of course, while the theorem itself stands unquestioned.

The role of proof in the process of acceptance is similar to its role in discovery. Mathematical ideas are discovered through an act of creation in which formal logic is not directly involved. They are not derived or deduced, but developed by a process in which their significance for the existing body of mathematics and their potential for future yield are recognized by informal intuition. While a proof is considered a prerequisite for the publication of a theorem, it need be neither rigorous nor complete. Indeed the surveyability of a proof, the holistic conveyance of its ideas in a way that makes them intelligible and convincing, is of much more importance than its formal adequacy (Hanna, 1983). Since fully adequate, step-by-step proof is in most cases impracticable, and since surveyability is lost when proofs become too long, proofs are conventionally elliptical and brief.

The conclusion therefore is that an orientation towards extreme formalism in proof is not reflective of current mathematical practice or current philosophies of mathematics.

There are, as has been shown above, good reasons for this. As Tymoczko has put it, "Mathematicians, even ideal mathematicians, are able to do mathematics and to know mathematics only by participating in a mathematical community."

5. CAREFUL REASONING

Despite the secondary nature of proof, it is easy to see how misunderstandings about the nature of mathematics arise. Mathematical results published for a mathematical audience are invariably presented in the form of theorems and proofs. They retain this form, reflecting as it does the nature of mathematics as a highly structured body of knowledge held together by the concept of logical precedence, even though the proofs are not judged by criteria of completeness or rigour. To a person only partially trained in mathematics, to someone who is neither fully equipped to assess significance nor able to make the intuitive judgments necessary in successfully surveying a proof, it might easily appear that the manner of presentation – with its possible implication that full rigour is the ideal form – is the core of mathematical practice. Thus competence in mathematics might readily be misperceived as synonymous with the ability to create the form, a rigorous proof.

It is only one step further, then, to assume that learning mathematics must involve training in the ability to create this form. To teach a beginning student is assumed to involve teaching the formalities of proof. Paradoxically, such an emphasis omits the crucial element. When a mathematician reads a proof, it is not the deductive scheme that commands most attention. It is, in fact, the mathematical ideas, whose relationships are illuminated by the proof in a new way, which appeal for understanding, and it is the intuitive bridging of the gaps in logic that forms the essential component of that understanding. When a mathematician evaluates an idea, it is significance that is sought, the purpose of the idea and its implications, not the formal adequacy of the logic in which it is couched.

It would therefore appear that what needs to be conveyed to students is the importance of careful reasoning and of building arguments that can be scrutinized and revised. While these skills may involve a degree of formalization, the emphasis must be clearly placed on the clarity of the ideas.

6. TEACHING

That reasoning is a pedagogical issue at all bespeaks a conviction that the learning of mathematics is a dynamic rather than static process, in which students progress towards deeper level of insight and skill. Thus a teaching activity that includes formal or informal reasoning can be judged to be of value only to the degree that it promotes greater understanding.

The starting point for understanding is the naive mathematical idea rooted in everyday experience. To provide a basis for further progress, this naive idea must be developed and made explicit. This requires a degree of formalism. A language must be created: symbols defined, rules of manipulation specified, the scope of mathematical operations delineated. Greater precision must be taught, so that the essential can be separated from the non-essential and greater generality achieved.

But this has its price. Distanced from the original intuitive context, the student may lose sight of reality and become a symbol pusher. Experienced mathematicians have learned to handle this danger by acquiring the ability to make mental shifts in moving among levels of generality and formalism, and by building on specific examples, drawing only upon those characteristics pertinent to the more general situation under study. They are able to exploit symbolism and algorithms to work automatically and efficiently, while retaining the ability to intervene in their own work to monitor its accuracy and effectiveness.

What are the issues to be kept in mind in teaching mathematics, then, and in particular in developing the power of reasoning?

1. Formalism should not be seen as a side issue, but as an important tool for clarification, validation and understanding. When a need for justification is felt, and when this need can be met with an appropriate degree of rigour, learning will be greatly enhanced.

2. It is not enough to provide mathematical experiences. It is the reflection on one's experiences which leads to growth. As long as students see mathematics as a black box for the instantaneous production of "answers", they will not develop the patience necessary to cope with the many and erratic paths their minds will take in trying to grasp what mathematics is about. One goal of pedagogy should be to help pupils maintain the level of concentration required to negotiate a line of reasoning.

3. Ironically for a discipline touted as precise, the student of mathematics has to develop a tolerance for ambiguity. Pedantry can be the enemy of insight. Sometimes an explanation is better given pictorially, loosely, by example or by analogy. Sometimes distinctions are better left blurred (e.g., the various roles of the minus sign and the use of "$f(x)$" as both the function and the value of the function at x). Sometimes the role of a symbol in the discussion should be allowed to vary (e.g., the parameter which is sometimes held constant, sometimes allowed to vary).

4. At the same time, when there is a danger that genuine confusion might develop, the student must learn to become conscious of looseness and to apply the necessary amount of rigour. It is this judgemental aspect of reasoning, so essential in mathematics education, that must be communicated to students.

II : COGNITIVE THEORY
OF
ADVANCED MATHEMATICAL THINKING

In this part of the book we begin to develop theories of cognitive development of particular value in advanced mathematical thinking. In Chapter One we singled out the importance of abstract definition and deduction at this level. In Chapter Two we saw how the processes of representation and abstraction play a crucial role. In Chapter Five Shlomo Vinner considers the differences between the abstract definition of a concept as given in a mathematical theory and the concept image as conceived in the mind of an individual. The research in the last decade clearly shows a wide gulf between desirable theory and actual practice. In Chapter Six, Guershon Harel and James Kaput consider the ways in which mathematical processes are encapsulated as conceptual entities and symbolized by notations in ways which may be more or less appropriate in different contexts. They too see that the formal definitions often play only a subsidiary role in mathematical thinking and continue the discussion of Dreyfus from Chapter Two by moving from mathematical processes to mental objects that can be manipulated. They reflect on the use of symbols in this thinking process and the manner in which the representation may be appropriately elaborated to enhance its meaning.

The encapsulation of a process as a mental object is subjected to deep analysis in Chapter Seven. Here Ed Dubinsky takes an in-depth look at the process of reflective abstraction, as originally conceived by Piaget for younger children, and extends Piaget's theories to advanced mathematics. He has a different emphasis from other authors in that he sees encapsulation of processes as objects as the main driving force in mathematical thinking and does not accept the prominent role given to visualization proposed in several other chapters. This difference exemplifies the divergence between two different kinds of mind cited from the observations of Poincaré in Chapter One. It is a fitting point to end the first half of our book.

CHAPTER 5

THE ROLE OF DEFINITIONS
IN THE TEACHING AND LEARNING OF MATHEMATICS

SHLOMO VINNER

1. DEFINITIONS IN MATHEMATICS
AND COMMON ASSUMPTIONS ABOUT PEDAGOGY

Definition creates a serious problem in mathematics learning. It represents, perhaps, more than anything else the conflict between the structure of mathematics, as conceived by professional mathematicians, and the cognitive processes of concept acquisition. It seems that no-one in the mathematical community disagrees with the claim that mathematics is a deductive theory and as such, it starts with primary notions and axioms. By means of the primary notions all other notions are defined. All the theorems, which are not axioms, are proved from the axioms by means of certain rules of inference. This might be a too short and oversimplified description, but essentially, it represents the common view of mathematicians about mathematics. It does not necessarily reflect the process by means of which mathematics is created, but it tends to be the way mathematics is presented in higher mathematics text books and mathematical periodicals. Of course, it is not possible to start with primary notions and axioms in every situation. Typically, one starts with well known notions and well known theorems and proceeds by defining new notions and by proving new theorems. This might have certain consequences for the way mathematics is taught, even before one starts to think about the appropriate pedagogy. Thus, mathematics teachers might form in their classes a sequence of definitions, theorems and proofs as a skeleton for their course. Following these consequences may be pedagogically wrong since the teaching should take into account the common psychological processes of concept acquisition and logical reasoning.

Let us describe some of the possible consequences which can be derived from considering the role of definition in mathematics. We claim that the presentation and the organization of mathematics in many text books and classrooms are partly based on the following assumptions:

1. *Concepts are mainly acquired by means of their definitions.*

2. *Students will use definitions* to solve problems and prove theorems when necessary from a mathematical point of view.

3. *Definitions should be minimal.* (By this we mean that definitions should not contain parts which can be mathematically inferred from other parts of the definitions. For instance, if one decides to define a rectangle in Euclidean geometry by means of its angles it is preferable to define it as a quadrilateral with 3 right angles and not as a quadrilateral having 4 right angles. This is because

65

in Euclidean geometry, if a quadrilateral has 3 right angles one can prove that its fourth angle is also a right angle.)

4. *It is desirable that definitions will be elegant*. For instance, some mathematicians think that the definition of the absolute value as $|x|=\sqrt{(x^2)}$ is more elegant than its definition as

$$|x| = \left\{ \begin{array}{ll} x; \text{ if } & x \geq 0, \\ -x; \text{ if } & x < 0. \end{array} \right.$$

Also, some mathematicians believe that the definition of a prime number (in the domain of whole numbers) as a number having exactly two different divisors is more elegant than its definition as a number greater than 1 divisible only by 1 and itself.

5. *Definitions are arbitrary*. Definitions are "man made". Defining in mathematics is giving a name. (For instance, when defining a trapezoid, one can define it as a quadrilateral having *at least* one pair of opposite sides which are parallel. On the other hand, he or she can define it, if they wish, as a quadrilateral having *exactly* one pair of opposite sides which are parallel. If you choose the first definition, a parallelogram is also a trapezoid. If you choose the second one, it is not. Now, if the idea that definitions are arbitrary is well understood the above fact will not cause a confusion, otherwise it might cause a great deal.)

The above five assumptions do not necessarily reflect all the aspects of definitions in higher mathematics. As claimed above, these assumptions are reflected very often in the pedagogy of teaching mathematics. A quick look at the majority of high school and college text books, and these demonstrate some concern to pedagogy, will show that definitions have major role in the presentation of course materials. Take for instance, the notion of absolute value of a number. Its best characterization is that it is the number without its sign or signs, This is quite clear to the students and this is what most of them tell you when you ask them about the absolute value. You can hardly find a text book which mentions it. Another possibility to characterize the absolute value of a number is to say that it is the distance of the number from zero on the number line. This is also quite clear to the students but perhaps less clear than the former characterization. You can find some text books and teachers who use it, but still the majority of teachers and text books will avoid it. So the majority of teachers and textbooks will use one of the definitions mentioned above. However, some teachers know that these definitions are quite unclear and confusing for most of the students. By advocating that it is possible not to use these formal definitions we do not ignore the need, at a *later* stage, to know that

$$|x| = \left\{ \begin{array}{ll} x; \text{ if } & x \geq 0, \\ -x; \text{ if } & x < 0. \end{array} \right.$$

The student should use it when solving algebraic equations and inequalities with absolute value. However, the above formula can be given and explained to the student at a later stage *as a claim* about the absolute value and not as its formal definition.

The point that we would like to make by discussing the example of the absolute value is the following: when coming to decide about the pedagogy of teaching mathematics one has to take into account not only the question how students are *expected* to acquire the mathematical concepts but also, and perhaps mainly, how students *really* acquire these concepts.

2. THE COGNITIVE SITUATION

"Against definition" is a title of a paper by Fodor *et al* (1980). The paper discusses the way that "the notion of definition has served to connect several aspects of classical theory of language with one another and with widely credited accounts of concept acquisition". The authors argue with some of the widely accepted views in cognitive psychology, especially with the following three:

1. "The definition of a word determines its extension" (p. 266).

2. "To understand a word is to recover its definition" (p. 274).

3. "Definitions express the decomposition of concepts into their elements" (p. 276).

Fodor *et al* claim that these views have no psychological ground. They bring some experimental evidence which disconfirms these views. Especially, according to Fodor *et al*, the following claim is refuted: "Understanding a sentence token involves recovering (i.e. displaying in working memory) the definition of such lexical items as the sentence contains". Thus, when understanding a sentence token, or when trying to understand it, people usually do not consult the definitions of the terms which occur in the sentence. Fodor *et al* deal with sentences taken from everyday life contexts. A careful examination of their claims, even without considering their experimental evidence, might lead to the conclusion that these claims are not only extremely reasonable but that they are even trivial. This is mainly because many words in everyday language do not have definitions (although they are "defined" somehow in dictionaries). Think of "car", "table", "house", "green", "nice", etc., and you realize immediately that when understanding, for instance, the sentence "my nice green car is parked in front of my house" you do not consult definitions. There is still the question what you do consult when understanding this sentence and we are not sure whether Fodor *et al*, give a clear answer to this question. With this particular sentence you will not consult definitions because there are no definitions for the words involved. On the other hand, contrary to everyday life contexts, there are the "technical contexts". In these contexts meaning is assigned to a term by a stipulation. Terms are defined as in mathematics. Hence, if you are in a "technical context" you should consult definitions, otherwise mistakes might occur. Of course, there is no need to consult definitions (which do not exist) when trying to understand the sentence "among all the cars at the parking lot my green car is the nicest". However, it is necessary to consult definitions when trying to understand the sentence: "among all rectangles with the same perimeter the square is the one which has the maximal area". Note that in everyday life contexts, a square is not considered as a rectangle by most of the people, whereas in all mathematical contexts a square is a rectangle.

When Fodor *et al* spoke "against definitions" they meant it in non-technical contexts. They wanted to refute a certain linguistic theory about the role of definitions in non-technical thought processes. However, in technical contexts, contrary to non-technical ones, the question is not how people behave but how they should behave. In technical contexts people are supposed to consult definitions of the technical terms involved. On the other hand, knowing the enormous impact that everyday life has on any situation, it will be reasonable to predict that definitions will be ignored by many people also in technical contexts. This really happens as we will show in the following. So, what do people consult when dealing with technical terms in technical situations? We will try to answer this question in the next section

3. CONCEPT IMAGE

A concept name when seen or when heard is a stimulus to our memory. Something is evoked by the concept name in our memory. Usually, it is not the concept definition, even in the case the concept does have a definition. It is what we call "concept image" (Tall & Vinner, 1981; Vinner, 1983) and others (Davis, 1984) call it "concept frame".

The concept image is something non-verbal associated in our mind with the concept name. It can be a visual representation of the concept in case the concept has visual representations; it also can be a collection of impressions or experiences. The visual representations, the mental pictures, the impressions and the experiences associated with the concept name can be translated into verbal forms. But it is important to remember that these verbal forms were not the first thing evoked in our memory. They came into being only at a later stage. For instance, when hearing the word "table", a picture of a certain table can be evoked in your mind. Experiences of sitting at a table, eating at a table, etc., can be evoked as well. You can recall that many tables are made of wood, most of them have four legs; usually you do not lie on a table, you can sit on a table but this can be regarded by some people as an impolite behavior. When you hear the word "function", on the other hand, you might recall the expression "$y = f(x)$", you might visualize a graph of a function, you might think of specific functions like $y = x^2$ or $y = \sin x$, $y = \ln x$, etc. From what we have said, it is clear that it is only possible to speak of a concept image in relation to a specific individual. In addition, the same individual might react differently to a certain term (concept name) in different situations. In Tall & Vinner (1981) the term "evoked concept image" is introduced to describe the part of the memory evoked in a given context. This is not necessarily all that a certain individual knows about a certain notion. In general, although we may not always use the term "evoked concept image" in what follows, the reader should always keep this in mind.

4. CONCEPT FORMATION

We assume that to acquire a concept means to form a concept image for it. To know by heart a concept definition does not guarantee understanding of the concept. To understand, so we believe, means to have a concept image. Certain meaning should be associated with the words. To know, for instance, that the power set of a given set is the set of all subsets of that given set, does not mean anything unless one can construct some power sets of given sets. Hence, the image of the power set concept might include some memories of the construction of some power sets.

Most concepts in everyday life, like house, orange, cat, etc., are acquired without any involvement of definitions. On the other hand, some concepts, even everyday life concepts, might be introduced by definitions. The word "forest" might be introduced to a child by saying "many, many trees together" (the Merriam Webster dictionary definition "a large thick growth of trees and underbrush" is, of course, a useless definition for a little child). Definitions like this help to form a concept image. But the moment the image is formed, the definition becomes dispensable. It will remain inactive or even be forgotten when handling statements about the concept in consideration. Thus, the "scaffolding metaphor" can be suggested for the role of definition, in concept formation: the moment a construction of a building is finished, the scaffolding is taken away.

5. TECHNICAL CONTEXTS

In technical contexts, definitions might have extremely important roles. Not only that they help forming the concept image but they very often have a crucial role in cognitive tasks. They have the potential of saving you from many traps which are set by the concept image. For instance, if you are asked to find a maximal value of a function in a closed interval and you recall a graph that corresponds to a local maximum and you try to differentiate the given function and to find the zeros of the derivative, then the explicit definition of a maximal value in a closed interval might help you to consider other possibilities different from local maximums. Sometimes, this can prevent mistakes. Not consulting the definition in the above case, might cause a fixation on the differentiating technique associated with the maximal value concept in the mind of many students. The differentiating technique leads to the desirable results in many cases but not in all.

Thus, technical contexts impose on students some thought habits which are totally different from those typical to everyday life contexts. One can predict that, at least in the beginning of the learning process, the thought habits of everyday life will take over the thought habits imposed by the technical contexts.

6. CONCEPT IMAGE AND CONCEPT DEFINITION – DESIRABLE THEORY AND PRACTICE

In order to present our ideas by means of diagrams (as in Vinner, 1983), assume the existence of two different "cells" in our cognitive structure (to avoid confusion, we do not mean biological cells). One cell is for the definition(s) of the concept and the second one is for the concept image. One cell or even both of them might be void. (The concept image cell is considered to be empty as long as some meaning is not associated with the concept name. This can happen in many situations where the concept definition is memorized in a meaningless way.) There might be some interaction between the two cells although they can be formed independently. A student might have a concept image of the notion of coordinate systems as a result of seeing many graphs in various situations. According to this concept image, the two axes of a coordinate system are perpendicular to each other. Later on, the student's mathematics teacher might define a coordinate system as any two intersecting straight lines. As a result of this, three scenarios might occur:

(I) The concept image may be changed to include also coordinate systems whose axes do not form a right angle. (This is satisfactory reconstruction or accommodation.)

(II) The concept image may remain as it is. The definition cell will contain the teacher's definition for a while but this definition will be forgotten or distorted after a short time, and when the student will be asked to define a coordinate system he or she will talk about axes forming a right angle. (In this case the formal definition has not been assimilated.)

(III) Both cells will remain as they are. The moment the student is asked to define a coordinate system he will repeat his or her teacher's definition, but in all other situations he or she will think of coordinate system as a configuration of two perpendicular axes.

A similar process might occur when a concept is first introduced by means of a definition. Here, the concept image cell is empty in the beginning. After several examples and explanations, it is gradually filled. However, it does not necessarily reflect all the aspects of the concept definition. Similar scenarios to (I)–(III) above might occur. This is shown in Figure 2.

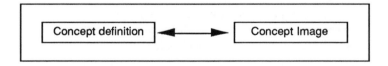

Figure 2 : Interplay between concept image and concept definition

Another illustration of (II) above is the following:

There are many students who are ready to swear that the definition of a limit of a sequence is a number to which the elements of a given sequence get closer and closer but never reach it. Thus the sequence whose nth element is given by $a_n=(-1)^{2n}$ does not have a limit (see also §7).

A further illustration of (II) above is the following:

Some students, after studying the modern concept of function, will say that a function is any correspondence between two sets which assigns to every element of the first set exactly one element of the second set. On the other hand they will not admit that the correspondence which assigns to every non-zero number its square and which assigns −1 to zero is a function (see also §7).

Figure 2 refers to the long term processes of concept formation. It seems to us that many teachers at the secondary and the collegiate levels expect a one way process for the concept formation as shown in figure 3, namely, they expect that the concept image will be formed by means of the concept definition and will be completely controlled by it.

Figure 3 :The cognitive growth of a formal concept

In addition to the process of the concept formation there are also the processes of problem solving or task performance. When a cognitive task is posed to a student the concept image and the concept definition cells are supposed to be activated. Again, it seems to us that many teachers at the secondary and the collegiate level expect that the intellectual processes involved with the performance of a given intellectual task should be schematically expressed by one of the three following figures (the figures represent only the aspect of concept image and concept definition involved in the process). The arrows in the figures represent different ways in which a cognitive system might function.

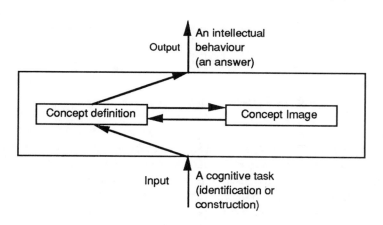

Figure 4 : Interplay between definition and image

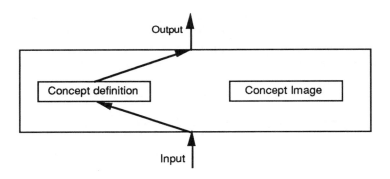

Figure 5 : Purely formal deduction

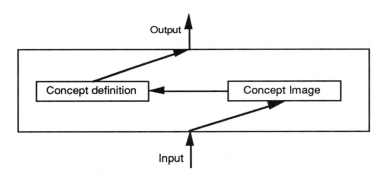

Figure 6 : Deduction following intuitive thought

The common feature of all the processes illustrated in Figures 4–6 is the following: no matter how your association system reacts when a problem is posed to you in a technical context, you are not supposed to formulate your solution before consulting the concept definition. This is, of course, the desirable process. Unfortunately, the practice is different. It is hard to train a cognitive system to act against its nature and to force it to consult definitions either when forming a concept image or when working on a cognitive task. Hence, a more appropriate model, for the processes which really occur in practice, is the following:

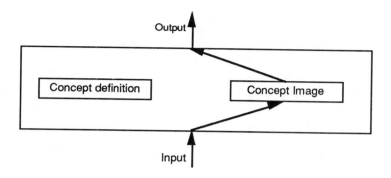

Figure 7 : Intuitive response

Here, the concept definition cell, even if non-void, is not consulted during the problem solving process. The everyday life thought habits take over and the respondent is unaware of the need to consult the formal definition. Needless to say, that in most of the cases, the reference to the concept image cell will be quite successful. This fact does not encourage people to refer to the concept definition cell. Only non-routine problems, in which incomplete concept images might be misleading, can encourage people to refer to the concept image. Such problems are rare and when given to students considered as unfair. Thus, there is no apparent force which can change the common thought habits which are, in principle, inappropriate for technical contexts.

Before closing this section we would like to remind the reader about the "evoked concept image" mentioned earlier in §3. In a specific cognitive task we deal only with one's evoked concept image. We do not claim that under different circumstances the same image will be evoked again. Thus, in our discussion, we do not evaluate somebody's cognitive system. Our analysis relates only to the part of the cognitive system which was activated when working on a given cognitive task.

7. THREE ILLUSTRATIONS OF COMMON CONCEPT IMAGES

In this section we will bring some experimental evidence to support our claim that the majority of the students do not use definitions when working on cognitive tasks in technical contexts. To be more specific, our claim is that the common high school and college courses do not develop in the science students, not majoring in mathematics, the thought habits needed for technical contexts. The students continue to use everyday life thought habits also in technical contexts. (Luckily enough for the students, this does not prevent them from passing exams.)

The concepts that we are going to discuss are the concept of function, the concept of tangent and the concept of limit of a sequence. Since a more detailed report about these can be found elsewhere (Davis & Vinner, 1986; Tall & Vinner, 1981; Vinner 1982, 1983) we will present here only the main aspects of the findings and the method of getting them. A natural method to learn about somebody's concept definition is by a direct question (what is a function? what is a tangent? and so on). This is because definitions are verbal and explicit. On the other hand, in order to learn about somebody's concept image usually indirect questions should be posed, as the concept image might be non-verbal and implicit. Thus the main task of the researcher is to invent questions that have the potential to expose the respondent's concept image. We will bring some of them. The following questionnaire was given to 147 students who studied mathematics at a high level in grades 10 and 11. In the first three questions the students were asked to choose between "yes" or "no" and to explain their answers.

1. Is there a function in which each number different from 0 corresponds to its square and 0 corresponds –1?

2. Is there a function in which each positive number corresponds 1, each negative number corresponds to –1, and 0 corresponds to 0?

3. Is there a function the graph of which is the following?

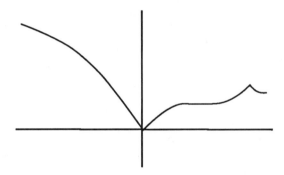

Figure 8 : Does this graph arise from a function?

4. In your opinion what is a function?

The concept of function was taught to all the students according to the modern approach, namely, a function is a correspondence between two sets which assigns to each element in the first set exactly one element in the second set. In spite of that, only 57% of the students gave this definition or something which is partly equivalent to it as an answer to question 4. (Note that we are dealing with good students. Thus, the figure 57% which can be considered as a great achievement in other circumstances is not so in this situation.) 14% of the students said that a function is a *rule* of correspondence and eliminated the possibility of an arbitrary correspondence. Rules cannot be arbitrary. They have to have a logical or mathematical ground. An additional 14% claimed that a function is an algebraic term, a formula, an equation or an arithmetical manipulation. The rest gave no answer or no satisfactory answer. When it came to concept images it turned out that at certain situations (questions 1 and 2) between one third to two thirds of the students think that a function should be given by one rule or, if two rules were given their domains should be half lines or intervals. A rule for a single point (like in question 1) is not permitted. Some students believe that correspondences which are not given by an algebraic rule are not functions unless the mathematical community has declared them as functions by giving them a name or a special notation. (This was reflected in answers to question 2). Other students (about 2/5) believe that a graph of a function should be regular, persistent, reasonably increasing etc. (This was reflected in answers to question 3.) Thus, many students who defined "function" correctly were not using their definition when replying to questions 1–3. In fact, only one third of the students who gave the correct definition of function also answered questions 1–3 correctly. No student with an incorrect definition answered questions 1–3 correctly.

Consider now the concept of tangent. It is usually introduced to mathematics students at a geometry course in the context of the circle. The definition of a tangent to a circle is an easy one and its visual representations is:

Figure 9 : A tangent to a circle

This picture can serve as a means to construct an image for the tangent concept in additional cases like:

Figure 10 : A mental image for a tangent

Students who take a calculus course usually get a formal or a semi-formal definition of the tangent to a graph of a differentiable function. However, their concept image, built up from experiences involving pictures like figures 9 and 10 may contain coercive elements which insist that a tangent may only meet the curve at one point and may not cross the curve at that point. As we shall see, such a concept image may lead the students to respond by drawing a line that is not a tangent at the required point, yet has the generic properties of the concept image. As in chapter 1, Tall (1987) termed such a concept a *generic* tangent.

The following questionnaire was given to 278 first year college students in calculus courses designed for science students (not majoring in mathematics).

> Here are three curves. On each one of them a point *P* is denoted. Next to each one of them there are three statements. Circle the statement which seems true to you and follow the instruction in the parentheses.
>
> A. Through *P* it is possible to draw exactly one tangent to the curve (draw it).
>
> B. Through *P* it is possible to draw more than one tangent (specify how many, one, two, three, infinitely many. Draw all of them in case their number is finite and some of them in case it is infinite).
>
> C. It is impossible to draw through *P* a tangent to the curve.

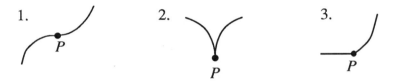

Figure 11: Which graphs have tangent(s) at *P* ?

> 4. What is the definition of the tangent as you remember it from this course or from previous courses. If you do not remember the definition of the tangent try to define it yourself.

The curves in 1,2,3 are the graphs of $y = x^3$, $y = \sqrt{|x|}$ and

$$y = \begin{cases} x^2; & x \geq 0 \\ 0; & x < 0 \end{cases}$$

but this was not given to the students. The tangent was defined in the above courses either as a limit of secants or as a line having a common point with the function graph whose slope is the derivative at this particular point. However, only 41% of the students gave one of the course definitions as an answer to question 4. 35% gave descriptions that suit the case of the tangent to a circle. They claim that a tangent touches the curve but does not intersect it, or that it meets the curve but does not cut it or that it has a common point with the curve but it is on one side of the curve. The rest gave no definition or meaningless definitions. The students concept images were expressed in the answers to questions 1, 2, 3 and are given in the following tables.

A	B	C	D	E
The right answer	A generic tangent	two tangents	Another drawing	No drawing
18%	38%	6%	10%	28%

Table I : Distribution of student drawings to question 1 (N=278)

A	B	C	D	E
The right answer	two tangents	infinitely many tangents	A 'balance' tangent	No drawing
8%	18%	18%	14%	42%

Table II : Distribution of student drawings to question 2 (N=278)

A	B	C	D	E	F
The right answer	A generic tangent	two tangents	Infinitely many tangents	Another drawing	No drawing
12%	33%	16%	7%	4%	27%

Table III : Distribution of student drawings to question 3 (N=278)

Some of the drawings are especially interesting. For instance, in 1B, 2B and 3B the students try to force the graph in order to meet the image formed by the tangent to the circle. 1B and 3B seem to be classic 'generic tangents' generated by their concept image, 2D is a generalization in which the 'tangent' is balanced on the cusp. In 1C, 2D (the bottom drawing) and 3C there is another phenomenon. It may be that the old concept image (a tangent to a circle) and the new concept image (constructed by the course definition) act at the student's mind simultaneously. It is a well known phenomenon in science learning, where, very often, old schemas are found together with new schemas in students' thinking.

Tall (1987) found some students responding with a dynamic image of a tangent, for example, intimating that the picture in figure 3B is such that the tangent "begins to turn" at the point in question, and so the tangent is drawn at a tilt to represent this tendency, even though the student concerned might sense that the turning does not actually begin until *after* the point concerned.

In 2C and 3D the students even invent the case of infinitely many tangents, on one hand, in order to meet the old image formed by the circle and on the other hand, realizing that there is no reason to prefer one "tangent", drawn according to the old image, to other, infinitely many, "tangents". Contrary to these students, there are the students in 2D (the top drawing) and 3B who, perhaps, prefer a kind of symmetry and thus stay with only one tangent, or perhaps, take as a starting point that there should be only one tangent and therefore conclude that it should be the one that has symmetry.

Finally, we will say few words about the notion of a limit of a sequence. Although our findings here come from a very small sample ($N=15$), they are more than typical because of the following reasons:

(1) The respondents are mathematically gifted students at a university high school.

(2) An "appropriate pedagogy" was used to teach them the notion of limit (with the teacher being aware of the necessity of bringing typical and non-typical examples of sequences which tend or do not tend to a limit. This is, of course, in addition to the formal definition. For more details see Davis & Vinner, 1986).

The concept was taught to the students at the end of their eleventh grade. Immediately after the summer vacation, on the first day of class, the following was given to the students by their teacher as a written test:

I need to know how much you remember about the concept of a limit of a sequence. Please, write a few paragraphs to show me what you remember. I suggested you may want to include:

1) A description of a sequence in intuitive or informal terms.

2) A precise formal definition.

Out of the 15 students only one gave an answer that can be considered as indication of reasonably deep understanding of the concept. This was:

The limit of a sequence is the number from which all the terms in the sequence, after a certain point, vary only by a little number ε.

(This answer misses the most important element of the formal definition, namely, a statement that the above is true for every $\varepsilon > 0$. Thus, this answer is treated quite literally. If only tough measures would have been taken then the result would be that not a single student showed a deep understanding of the formal definition. The ability to construct a formal definition is for us a possible indication of deep understanding. Of course, it is not sufficient, since the reconstruction of a formal definition can be obtained by rote memorization).

In the other 14 students some typical misconceptions were found which influenced the formal definitions that the students were asked to give. In our terminology, the concept definition was reconstructed by referring to the concept image. Since the concept image was incorrect this resulted in an incorrect formal definition. The main misconceptions were:

(1) A sequence "must not reach its limit" (thus, the sequence: 1,1,1,... would be said not to converge to a limit),

(2) The sequence should be either monotonically increasing or monotonically decreasing (thus, for instance the sequence whose nth element is given by $a_n = 1+(-1)^n/n$ does not tend to a limit),

(3) The limit is the "last" term of the sequence. You arrived to the limit after "going through" infinitely many elements.

In the three central concepts discussed above there is a conflict between the formal definition and the concept typical examples which might cause an incorrect concept image. The findings show that, in spite of the emphasis which was given to the definition of the concepts, many students did not use them when working on tasks in which formal definitions should have been used. This can lead to two opposite conclusions:

(l) Giving up the goal of changing the students' thought habits from the everyday mode to the technical mode.

(2) Trying to change the students' thought habits by an appropriate treatment (perhaps as an independent topic which might lead to more awareness. The integration of this topic in the common courses does not attract enough attention that can lead to the desired results). More about this dilemma in the next section.

8. SOME IMPLICATIONS FOR TEACHING

We would like to recommend here two didactical rules which are relevant to the problem raised in this paper.

(1) To avoid *unnecessary* cognitive conflicts with students,

(2) To initiate cognitive conflicts with students when these conflicts are necessary in order to enhance the students to a higher intellectual stage. (This should be done only when the chance of reaching a higher intellectual stage is reasonably high.)

We claim earlier that one of the goals of teaching mathematics should be changing the thought habits from the everyday life mode to the technical mode. This cannot be done in a short period and cannot be successful with everybody. Our belief is that mathematical concepts, if their nature allows it, should be acquired in the everyday life mode of concept formation and not in the technical mode. One should start with various examples and non-examples by means of which the concept image will be formed.

This does not mean that the formal definition should not be introduced to the student. However, the teacher or the text book writer should be aware of the effect that such introduction can have on the student's thinking. (If the concept is not too complicated the teacher can even ask the students to suggest *their* definition for the concept.) If our students are candidates for advanced mathematics then, no doubt, they should be trained to use the definition as an ultimate criterion in various mathematical tasks. But in order to achieve this goal, one should do more than introducing the definition. One should point at the conflicts between the concept image and the formal definition and deeply discuss the weird examples (like the tangent to the graph of $y = x^3$ at (0,0) or the limit of the sequence whose nth element is $(-1)^{2n}$, $n=1,2,3...$, etc.). If, on the other hand, our students are not candidates for advanced mathematics, then it is better to avoid the conflicts. There is no harm if the students memorize the formal definition and repeat it in various occasions. The teacher and the text book writer, on the other hand, may even feel that they have completed their task by introducing the formal definition. But they should have no illusions about the cognitive power that this definition has on the student's mathematical thinking.

Thus the role of definition in a given mathematics course should be determined according to the desired educational goals supposed to be achieved with the given students. If the students are candidates for advanced mathematics then, not only that definitions should be given and discussed, the students should be trained to use them as an ultimate criterion in mathematical tasks. This goal can be achieved only if the students are given tasks that cannot be solved correctly by referring only to the concept image. As long as referring to the concept image will result in a correct solution, the student will keep referring to the concept image since this strategy is simple and natural. Only a failure may convince the student that he or she has to use the concept definition as an ultimate criterion for behavior. Thus we do believe that changing students' thought habits from the every day mode to the technical mode is an important goal for teaching mathematics. Contrary to Fodor *et al* (1980) we campaign *for* definition and not *against* definition but we claim that this aspect of definition cannot be achieved with all students. There might be various opinions about the percentage of students who are capable of this aspect and there is also the practical question how to decide whether a certain student can change his thought habits from the everyday life mode to the technical mode. We do not have answers to these questions yet. Therefore, the decisions about the goals of teaching definitions should be left to the intelligent and sensitive mathematics teacher.

The role of definition in mathematical thinking is somehow neglected in official contexts (text books, documents about goals of teaching mathematics, etc.). We are not sure whether this is because it is taken for granted or because it is overlooked. It is obligatory to remember that there are some contexts in which referring to the formal definition is critical for a correct performance on given tasks (among them there are the identification of examples and non-examples of a given concept, problem solving and mathematical proofs). On the other hand we want to be realistic about the chance of achieving the above goals. We do not believe in "mathematics for all". We do believe in some mathematics for some students. And even this can be achieved only by appropriate pedagogy under appropriate conditions for learning.

CHAPTER 6

THE ROLE OF CONCEPTUAL ENTITIES
AND THEIR SYMBOLS
IN BUILDING ADVANCED MATHEMATICAL CONCEPTS

GUERSHON HAREL & JAMES KAPUT

Mathematical thinking is carried out using mental objects. For example, suppose one asks if a vector space V and its double dual V^{**} are isomorphic. At one level, one is asking about the "objects" V and V^{**} and, to begin describing an isomorphism, one may go on to describe a correspondence between respective vectors in the two spaces, which again, are treated mentally as objects, although they might be n-tuples or matrices, for example. Similarly, one may need to define a mapping between two function spaces, where the elements of the domain and range of the mapping must be treated cognitively as objects, as opposed to the mapping itself, which may be treated as a process, with inputs and outputs. In yet another instance, one may need to reinterpret a universal construction in the sense of MacLane (1971) as an adjoint functor pair, where the existence of a unique mapping with a certain property in fact defines a natural transformation between functors – so the mapping must play the role of an object on which the natural transformation acts. Such experiences are quite common in mathematics at all levels, but they feature widely throughout advanced mathematical thinking. The aim of this chapter is to begin to discuss them and their roles in helping us to build ever more complex mathematical concepts.

The idea of conceptual entities formation was suggested by Piaget (1977) in his distinction between form and content. Recently, several researchers have recognized its value in the learning of mathematics. It has been called *encapsulation* (Ayers, Davis, Dubinsky & Lewin, 1988), *reification* (Sfard, 1989), *integration operation* (Steffe & Cobb, 1988), for example, this process is an instance of reflective abstraction (Beth & Piaget, 1966), in which "a physical or mental action is reconstructed and reorganized on a higher plane of thought and so comes to be understood be the knower" (p. 247). Greeno (1983) defines a conceptual entity as a cognitive object for which the mental system has procedures that can take that object as an argument, as an input. He distinguishes cognitive objects from attributes, operations and relations, which attach to or act on objects. Further, he suggests that to qualify as objects, they must be permanently available in the individual's mental representation (p. 277).

The construction of function as a conceptual entity is an example of the entification process (Thompson, 1985a; Harel, 1985; Ayers *et al*, 1988). One level of understanding the concept of function is to think of a function as a *process* associating elements in a domain with elements in a range. This level of understanding may be sufficient to deal with certain situations, such as interpreting graphs of functions point-wise or solving for x in an equation of the form f(x)=b, but it would *not* be sufficient to deal meaningfully with situations which involve certain operators on functions, such as the integral and differential operators, as we will see later in this chapter. For the latter situations, the three components of function – the rule, the domain, and the range – must be encapsulated into a single conceptual entity so that these operators can be considered as procedures that take functions as arguments. Incidentally, a formal definition of a function as a single *set* of ordered pairs, a mathematical entity, does not appear to play a role in these situations – when would one conceive of a function as a set of ordered pairs in the context of applying a differential operator to that function? In this way the concept image evoked in a given context may be different from the formal definition, and may even at times be in conflict with that definition, as discussed in the previous chapter.

The construction of conceptual entities embodies the "vertical" growth of mathematical knowledge (in the sense of Kaput, 1987). For example, at lower levels, the act of counting leads to (whole) numbers as objects, taking part-of leads to fraction numbers, functions as rules for transforming objects become themselves objects that can then be further operated upon, for instance they may be differentiated or integrated. This complements the kind of "horizontal" growth associated with the translation of mathematical ideas across representation systems and between non-mathematical situations and their mathematical models.

In the next section of this chapter we lay out some of the circumstances under which conceptual entities are created and used and what their cognitive function might be, often by pointing to consequences in students' reasoning processes where they have not yet been mentally constructed. In the following part we will shift attention to the complex roles of notation systems in building and using conceptual entities. We regard this chapter as a foray into relatively unexplored territory, and do not make claims of completeness or of empirical substantiation for the framework being suggested.

1. THREE ROLES OF CONCEPTUAL ENTITIES

We will discuss the concepts of function, operator, vector space, and limit in terms of the role that conceptual entities have for:

1. *Alleviating working memory or processing load* when concepts involve multiple constituent elements.

2. Facilitating *comprehension* of complex concepts: the cases of "uniform" operators, "point-wise" operators, and "object-valued" operators.

3. Assisting with the *focus of attention* on appropriate structure in problem solving.

Greeno (1983) suggested a number of functions of representational knowledge involving conceptual entities: forming analogies between domains, reasoning with general methods, providing computational efficiency, and facilitating planning. He offered empirical findings that are consistent with his suggestions; these findings deal with elementary mathematics – geometry proofs and multi-digit subtraction – as well as physics, puzzle problems, and binomial probability. He also suggests that instructional activities with concrete manipulatives can lead to an acquisition of representational knowledge that includes conceptual entities. Other researchers suggest different types of instructional activities for the construction of conceptual entities. For example, Ayers *et al*, (1988) demonstrate how computer activities in learning mathematical induction and composition of functions can facilitate the construction of these concepts as entities (see the next chapter).

1.1 WORKING-MEMORY LOAD

One psychological justification for forming conceptual entities lies in their role in consolidating or chunking knowledge to compensate for the mind's limited processing capacity, especially with respect to working memory. To avoid loss of information during working memory processes, large units of information must be chunked into single units, or conceptual entities. Thus, thinking of a function as a process would require more working-memory space than if it is encoded as a single object. As a result, complex concepts that involve two or more functions would be more difficult to retrieve, process, or store if the concept of function is viewed as a process. This is true for many concepts in advanced mathematics. Imagine, for example, the working-memory strain in dealing with the concept of the double dual space of a space of $n \times n$ matrices if none or only a few of the concepts, matrix, vector space, functional, and field are conceived as consolidated entities.

1.2a COMPREHENSION: THE CASE OF "UNIFORM" AND "POINT-WISE" OPERATORS

Despite the heavy working-memory load involved in understanding the dual space of an $n \times n$ matrix space without most of its subconcepts being entities, it is still possible to make sense out of it, at least momentarily. In some situations, however, the justification for the formation of conceptual entities is more than just a matter of cognitive strain that results from a memory load. In such situations comprehension *requires* that certain concepts act mentally as objects due to an intrinsic characteristic of the construct involved. Examples of such situations include those which involve the integral or differential operators. These types of "uniform" operators *cannot* be understood unless the concept of function is conceived as a total entity. We distinguish these from other types of operators on functions which could be termed "point-wise" operators, and for which there is no need to conceive functions as objects, but only as processes acting on individual elements of their domains. For example, sum and composition can be treated as point-wise operators; this position is different from Ayers *et al*'s (1988) position who argue that composition of functions requires the encapsulation of function as an entity. Further research is needed to examine the two arguments. The cognitive process of understanding these operators involves the conception of a function as a process acting on individual elements of the domain. In constructing the composition of two functions f and g, say f∘g, one must first perform the process g on an arbitrary element x of the domain, generating a result $g(x)$, and then perform the process of f on that result to obtain $f(g(x))$, all conceivable as acting on individual elements of the domain. These two separate operations are coordinated to produced a new process. Similarly, in constructing f+g, for every input x, the outputs, $f(x)$ and $g(x)$, are produced to construct the sum, $f(x)+g(x)$. This sum can even be illustrated graphically by using a sample set of directed line segments for the distances between the horizontal axis and the graphs of f and g, respectively. Then the graph of f+g is the graph whose distance from the horizontal axis is given by the vector sum of the directed line segments. Clearly, the sum f+g can be illustrated point-wise.

The limit of a one variable function is another case which may be regarded as a point-wise operator. To understand this complex concept, many clusters of knowledge about different concepts in mathematics are required whose rich conceptual content is reflected in the complexity of its historical development. We will not attempt to analyze this knowledge here; however, the process-conception of function is sufficient (and necessary) to understand the limit concept. This is so because $\lim_{x \to a} f(x) = L$ may be viewed in terms of the *point-wise dependency* between the behavior of the numbers "near" a, x's, or inputs of f, and the behavior of their outputs, $f(x)$'s, "near" L.

By contrast, "uniform" operators arise when the point-by-point process is *inapplicable*. For example, to understand the meaning of:

$$I(t) = \int^{t} f(x)\, dx$$

as a function of t, it is necessary to think of $I(t)$ as an operator that acts on the process $x \to f(x)$ *as a whole* to produce a new process:

$$t \mapsto \int^{t} f(x)\, dx$$

It is the awareness of acting on a process as a whole, as a totality – not point-by-point – that constitutes the conception of that process as an object.

Mathematically unsophisticated students attempt to interpret "uniform" operators as "point-wise" operators apparently because they cannot conceive of a function as an object. Consider the derivative operator. Our experience in the classroom suggests that many students understand that $f'(x)$ means: for the input x there is the output $f(x)$, and for that output we get the derivative $f'(x)$. Faced with the question,

find the derivative of the function $f(x) = \begin{cases} \sin x & \text{if } x \neq 0, \\ 1 & \text{if } x = 0, \end{cases}$

a common response is:

$$f'(x) = \begin{cases} \cos x & \text{if } x \neq 0, \\ 0 & \text{if } x = 0. \end{cases}$$

The student is no longer treating differentiation as a limit process, but as an algorithm to be applied to the formula at each point (or to the two separate formulas in the expression). To be able to handle this problem, the student needs to be able to consider the values of the function near x and renegotiate the limit process. In Greeno's terms, the function f must act as an argument for the (cognitive) differentiation operator, which it cannot do unless the function is conceived as a conceptual entity.

1.2b COMPREHENSION: THE CASE OF OBJECT-VALUED OPERATORS

As the notion of function develops, it can have different objects as inputs and outputs, in particular, it can output another function. For instance, the real-valued function $f(x,y)$ is usually thought of as a process mapping points on the plane, (x, y), into points on the real line, $f(x,y)$; thus, students who possess the process-conception of function would likely have no difficulty dealing with this interpretation. A more subtle interpretation can view $f(x,y)$ as a process which associates points on the real line, x, with functions, $f_x(y)$ where the latter assigns the value $f(x,y)$ to y. In this interpretation f is regarded as a function with input x and output the function f_x. We believe that, cognitively, thinking of a function as an output is *not* different from thinking of it as an input, in the sense that in both cases a function must be treated as a variable, as a conceptual entity. In this respect, this interpretation of $f(x,y)$, like the "uniform" operator, demands that the concept of function will be treated as an object. However, the cognitive demands of such a viewpoint are often great.

This analysis, which has yet to be empirically substantiated, is supported by our informal observations while teaching undergraduate mathematics classes the concepts of double limit, $\lim_{(x,y) \to (a,b)} f(x,y)$, and the iterated limit, $\lim_{x \to a} \lim_{y \to b} f(x,y)$. As some textbook authors have indicated (e.g., Munroe, 1965, p. 108), we observed that while computationally the iterated limit is easier than the double limit, conceptually the iterated limit involves a more sophisticated idea, which causes difficulty for students in particular circumstances. In stating and proving certain theorems on iterated limits (e.g., theorems concerning conditions on equality between this limit and the double limit), one needs to regard $\lim_{x \to a} \lim_{y \to b} f(x,y)$ as a composition of the following three mappings (see figure 12):

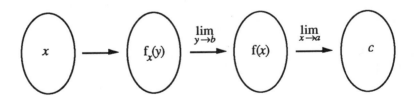

Figure 12 : $\lim_{x \to a} \lim_{y \to b} f(x,y)$ as a composition of three mappings

1. $M: x \to f_x(y)$, whose domain is a set of real numbers and whose range is a set of functions;

2. $\lim_{x \to a} : f_x(y) \to f(x)$, whose domain and range are sets of functions;

3. $\lim_{x \to a} : f(x) \to c$, whose domain is a space of functions and range is a set of numbers.

Students responses and questions indicate difficulty in dealing with aspects concerning the operator M, which, as indicated earlier, requires the object-conception of function. While the operator M must be understood as an object-valued operator, the other two operators, $\lim_{y \to b}$ and $\lim_{x \to a}$ can be viewed in two ways, which determine different levels of understanding the concept of iterated limit. In one way $\lim_{y \to b}$ and $\lim_{x \to a}$ are uniform operators acting on objects which happen to be functions. This level of understanding, although desirable, is not achieved by the average student, who usually views these limits, and the concept of limit in general, in a less sophisticated way as point-wise operators.

Besides the iterated limit, the undergraduate mathematics curriculum is replete with situations involving object-valued operators, for example those which concern parametric functions, such as $f(x)=ax+b$, $f(x)=\sin(ax)$, $f(x)=\log_a x$, etc., or parametric equations involving such functions. In these situations the correspondence between the parameters and the function, or the equation, constitutes an object-valued operator. The difficulties involved in understanding object-valued operators was investigated by Harel (1985) in the context of linear algebra (taught to advanced high-school students in Israel). It was found that students usually had difficulty dealing with such a correspondence, unless they were able to tag the outputs of the correspondence with familiar geometric figures, such as lines or planes (e.g., $t \to (a, b)+t(c, d)$ or $(t_1, t_2) \to (a,b) + t_1(c, d) + t_2(e,f)$). These geometric figures, which were manipulable objects for the students, apparently helped the students to construct such a correspondence as an object-valued operator.

Another common example involves the construction in abstract algebra of the quotient object associated with a "normal" sub-object, e.g., in the case of groups. The cosets must be conceived as objects if they are to participate as elements of a group. However, the existence of a "representative element" for a coset, where the operation defined on cosets can be given in terms of an operation on their representatives, makes it possible to deal successfully with many aspects of the quotient group on a symbol manipulation level without treating the subsets of the group as objects, or even as subsets. Students' inadequate conceptions are revealed when one asks them to attempt to create a group using a non-normal subgroup's cosets – they often cannot understand why the subsets "fall apart" when they attempt to multiply them together as sets, or by using representatives.

Finally, data reported by Kaput (in press) can further support the cognitive distinctions among the different types of operators made above. Secondary level students were asked to determine an algebraic rule that fits a student-controllable set of numerical domain-data (they pick the x's and the computer provides the $f(x)$'s). Examination of their behavior revealed a clear and stable decomposition of the group of students (in a sample of over 40 high school students) into two sets, one of whom consistently used a point-by-point pattern-matching process, mediated by natural language formulations of their proposed "rules," while the other searched for and applied a parametrically mediated formulation of their proposed rules. The latter, for example, would look for constant change in the dependent variable, identify this as the "m" in $y=mx+b$, and proceed from there. For them the process was a search for parameters that indexed functions as objects. In effect, they were dealing with a space of functions (albeit a limited one), whereas the other group of students conceptualized the task as a point-wise attempt to build a function whose point-wise behavior matched the rule that they had formulated using natural language.

1.3 CONCEPTUAL ENTITIES AS AIDS TO FOCUS

The third role of conceptual entities we have identified involves facilitating focus on those aspects of a problem representation that are most relevant to the solution of a problem. In a one-on-one interview with an experimental group of Israeli high-school students regarding the concept of vector space (after several instructional sessions in which this concept was gradually abstracted from two and three dimensional representations; see Harel, 1989a, 1989b), the first author asked the following question:

> Let V be a subspace of a vector space U, and let β be a vector in U but not in V.
> Is the set $V+\beta=\{v+\beta \mid v$ is a vector in $V\}$ a vector space?

There were clearly two groups of students: those who answered this question by checking the whole list of the vector-space axioms, and those whose answer was something like, "you moved the whole thing, it doesn't have the zero vector any more", or "the new thing, $V+\beta$, is not closed under addition". Clearly, the latter group of students viewed V as a total entity, a "thing," and thus they were able to view $+\beta$ as a shift operator which takes V as an argument, an input. This enabled them to focus on those vector space properties that are most relevant to the solution of the given problem, namely, the zero property or one of the closure properties. The other group of students, on the other hand, relied on the formal definition of vector space by checking whether the individual axioms apply. That $V+\beta$ is a subset of the vector space U, which guarantees the existence of most of the axioms, was not visible to these students. Moreover, many of these students failed to check some of the axioms, including those essential to the solution of the problem (e.g., the existence of zero).

2. ROLES OF MATHEMATICAL NOTATIONS

The power of mathematics associated with the roles of conceptual entities is closely related to the roles of mathematical symbolism. Using mathematical notations, complex ideas or mental processes can be chunked and thus represented by physical notations which, in turn, can be reflected on or manipulated to generate new ideas. In this section we will discuss three aspects of the interaction between formation of conceptual entities and mathematical notation:

1. The role of mathematical notation in forming conceptual entities.

2. Different types of mathematical notations, *elaborated* and *tacit* notations, and the manner in which they represent conceptual structure.

3. Notations as substitutes for concepts.

2.1 NOTATION AND FORMATION OF COGNITIVE ENTITIES

Greeno (1983) stated two conditions that help distinguish entities from other mental events. One is its continual presence in a mental representation; the other (mentioned earlier) is its ability to act as an argument in another mental procedure or argument. By providing continual *perceptual* experience, material notations help provide the basis for continuing *conceptual* presence. This role is based simply on notations as names – the notation serves to *name* an item in our conceptual world. We might term this the "nominal" role. Note that the parts of the syntax of a notation system associated with identification and discrimination of notational objects plays an important role here. Having an explicit name for a mental event helps objectify it through a kind of transference of object permanence – from the permanence of the physical notational name (which produces perceptual experience on a more or less continuous basis) to a cognitive permanence. Of course, the perceptual item must somehow come to be integrated with the conceptual one. Otherwise, all one might end up with is, say, an easily reproducible mental experience of a mark or character string, with no other mental activity or structure beyond that primitive experience – which is the experience of altogether too many students.

The nominal role of symbols is frequently played out using conventions that help distinguish the status or differing roles of objects in complex situations – convention-based variations in the names of objects help distinguish the classes to which they belong. Suppose a concept involves a process which takes entities of a different order as inputs and outputs, e.g. differentiation operating on functions. Then there is a need to distinguish between the higher level process and its lower level inputs and outputs, a need which is typically satisfied by using systematically different symbols for the items at each level. Then the conceptual activity of keeping the things distinguished is off-loaded onto the notation system. For example, many higher level mathematical activities involve defining functions between sets of functions – as between a vector space and its double dual. Another typical example occurs in topology, when one defines various compactifications, e.g., the Stone-Czech compactification of a regular Hausdorff space based on sets of continuous functions on the unit interval. In all such cases, one finds that, typically, different classes of characters are used to distinguish the different levels of functions - say, one Greek and the other contemporary English-based.

Systematic variation in names also is employed through the use of different classes of symbols to distinguish when an object is being treated in two different ways, where it has essentially two different identities. Consider the conventions used to distinguish the identity of a real number x from its identity as a member of the field of complex numbers, where it may be denoted by $x+0i$. Similar distinctions are made whenever a canonical embedding is being employed, not merely in the case of algebraic closures, because it is a characteristic of "canonicalness" that the substructure is maintained within the larger structure. A related case involves the distinction between a constant function and its value. In all these cases, object identity is identified and maintained notationally.

Relative to Greeno's second condition for cognitive entities acting as arguments in other procedures, the syntax of a notation system specifically structures the place of the material notational objects in a coherently organized physical system. Such a system is designed to support a given type of thinking. For example, the character string notation for functions supports highly sophisticated manipulations, which in turn, are used to facilitate a wide variety of mental operations on the conceptual objects that those character strings denote. Thus, the act of factoring the character-string representation of a polynomial function to help identify its roots may be based on some syntactic rule (e.g., applied to the difference of cubics), which obviates the need to justify all the steps of the process. The strength of a notation system may be measured by whether, and to what degree of fidelity, syntactically guided actions on its objects reflect and/or subsume important mental operations.

We conclude this section with two specific examples to illustrate the variety of ways notations either help encapsulate mathematical concepts as entities or supplant conceptual entities in reasoning processes. Goldin (1982) discusses the impacts of languages or notations on the different stages of the problem solving process, citing his own data as well as the well-known problem-isomorph work by Simon and colleagues. The following discussion can be thought of as somewhat preliminary to the issues discussed by these researchers in the sense that we are dealing with the concept-notation relationship at a more primitive level.

Example 1: Consider the use of graphical notation, the slope of straight lines, to facilitate the order comparison between ratios described as linear functions between sets of objects, measures, or even numbers. To compare two such on the basis of a table of data (a sequence of ordered pairs) or even on the basis of a pair of fractions is not as easy as comparing the slopes of their associated straight lines in a coordinate plane. In this case one need only attend to two things (2 lines) as distinguished by their most salient attribute, their slope. Each single line embodies an infinite set of equivalent pairs of ratio values. This seems to be an instance of a one-for-many substitution of a single notational object for a set of mental objects, although from another perspective it amounts to an integration of detailed features into a single object.

Example 2: Recall the study mentioned in §1.2 where students were determining functions from numerical data. There were two types of students: One type of students were essentially "pre-algebraic" in their thinking, and treated every potential rule that they inferred from the numerical data in a table (which they generated) as a natural language-based rule. That is, they thought of $2x+1$ as doubling and adding one, in terms of a natural language interpretation, rather than in terms of parameters m and b in $mx+b$. Thus they did not see growth in the numerical data in the same way as those who were looking for values of these parameters. Basically, the latter were looking for growth rates, which they interpreted as the first parameter's value, etc. For them, a linear function was experienced as a "thing", a conceptual entity, whose identity is determined by the two parameters. The other students were looking for a way to translate from their natural language-based encoding of an unencapsulated *process* to algebra. They quite often succeeded – as long as the parameters involved were positive whole numbers. For negatives, they fell apart, because they were not able to get easy natural language encodings of what for them was a process rather than a thing (Kaput, in press). An open question is what is the relation between the conceptual entity and the parameter notation? Which came first? Or did they co-evolve? In any case, this example seems to offer an instance of the functional power of the nominal use of symbols – as do most systematic uses of parameters.

2.2 REFLECTING STRUCTURE IN ELABORATED NOTATIONS

The inventors of mathematical notations created them to express the contents of their own minds, both to themselves, to aid their own thinking, and to others, to aid in the communication of their conceptions. As Leibniz, that great master of notation-invention put it,

In signs one observes an advantage in discovery which is greatest when they express the exact nature of a thing briefly, and, as it were, picture it; then indeed the labor of thought is wonderfully diminished.   (Quoted in Cajori, 1929, p.184)

Extending his remark, we might add that the structure of the conceptions is, in some way, being reflected in the structure of the notations, especially in their syntax. Or, put more constructively, the experience of perceiving the notations shares important features with the experience of the conception apart from any perceptual act. Extending this observation further, we suggest that it is even more important that actions on notational objects in some regular way reflect mental actions on the conceptions. (We again hasten to add, however, that we are not suggesting any kind of simple relationship between notation and conception!)

But mathematical symbols differ in the extent to which they include features that reflect the structure of the mathematical objects, relations or operations that they stand for. Some are more elaborated than others (Harel, 1987). For example, the place-valued symbol 324 expresses a specific structure of the quantity it represents: three hundreds, two tens, and four ones. Of course, this number written in expanded notation is even more elaborated. Similarly, the more abstract symbols, (x, y) for an ordered pair of numbers, $f(x) = 3x^2$ for a specific real-valued function, AB for a line segment whose endpoints are A and B, and

$$\begin{pmatrix} a_{11} & a_{12} & \cdots & a_{1n} \\ a_{21} & a_{22} & \cdots & a_{2n} \\ & & \cdot & \\ & & \cdot & \\ & & \cdot & \\ a_{m1} & a_{m2} & \cdots & a_{mn} \end{pmatrix}$$

for an $m \times n$ matrix, are all relatively *elaborated* symbols, because they encode the structures or relationships among components of their referents.

On the other hand, for example, the concept "matrix of the linear transformation T relative to the pair of ordered bases μ and ν" can be symbolized by the significantly less elaborated symbol $[T]_{\mu,\nu}$. A more elaborated symbol for this concept could be $[T]_{\mu \to \nu}$, which indicates that the matrix representation of a transformation T depends on the relationship between the bases in its range and in its domain. An even further elaborated symbol for this concept is:

$[[T(\mu_1)]_V: [T(\mu_2)]_V: \ldots : [T(\mu_n)]_V]$ (used by Anton, 1981),

which encodes many of the variables included and its referent. In contrast, the symbol $[t_{ij}]$ (used by Nering, 1970, to represent the same concept) is far less elaborated, whilst the bare symbol T is a non-elaborated, or *tacit*, symbol. Tacit symbols provide essentially an indexical function – they *name* things, without denoting aspects of the structure of what is named.

One category of tacit symbols consists of those which, during a discussion or proof, are used to represent variables. For example, the statement, "let β be an ordered pair ...", typifies a context in which such a tacit symbol is used; here, the symbol β does not encode the structure of its referent – an ordered sequence of two objects – but it, together with the surrounding phrases, *does* name the set over which the variable varies.

The extent to which a notation is elaborated is determined by the extent to which it ties to prior mathematical knowledge, which is very much a cognitive matter. Indeed, what is elaborated for one person may appear very bare and tacit for another. Nonetheless, the act of connecting a bare notation to an elaborated one is a translation act, which, depending on circumstances, may operate in either direction. The notation's perceived connection with prior knowledge takes the form of perceived features that reflect features of the prior knowledge. For example, two different symbols are usually used to represent the composition of two functions f and g: $f(g(x))$ and $(f \circ g)(x)$. The symbol $f(g(x))$ expresses the process in which the two functions are composed: the input x in the function-machine g produces the output $g(x)$, where $g(x)$ now acts as an input in the function-machine f to produce the output $f(g(x))$. (Note the strong use of temporality here.) Thus the symbol $f(g(x))$ is amenable to the thinking of a function as a process, but depends on the prior knowledge of input-output relations expressed using the standard $f(x)$ notation. The symbol $(f \circ g)(x)$, on the other hand, describes an operation between two functions – f and g – which produces a third one – $(f \circ g)(x)$. This symbol describes f and g as inputs in the [meta] function-machine $\circ$, and thus to understand its meaning functions must be viewed as conceptual entities. In this example, the prior knowledge is that of operating on inputs to functions, and the notation feature is reflected in a parallelism of structure, except that the first function in the composition acts as the input.

The pedagogical importance of this example is that some mathematical symbols cannot be understood via the symbol $f(g(x))$; for example, the "uniform" operator:

$$I(x) = \int^x f(t)\, dt.$$

Students have trouble thinking of the integral as a function of x – which is revealed when they are asked to treat it like a function. Our notation $I(x)$ for it is itself intended to help with this – it assists entification by treating it notationally as a function, elaborating it in such a way that the functional dependence on the variable x is highlighted.

The distinction between elaborated symbols and tacit symbols has important consequences for learnability and usability. In Harel (1987) it was hypothesized that an elaborated symbol would be better understood and remembered if it expresses the main and salient variables in its referent. Here, we additionally hypothesize that a tacit symbol can be more meaningfully used when its referent is encapsulated into a conceptual entity. That is, in developing a symbol for a concept one must try to match the degree of elaboration of the symbol with the degree of elaboration of the user's concept, which in turn must match the user's needs for the task at hand. After all, in some cases it is important to suppress detail, and in others the detailed structure plays a role in what one is trying to do. It seems, then, that one's control of the amount of structure explicitly represented in the symbolism is a major factor in mathematical thinking, because one can adjust the "focus of one's mental microscope" by adjusting the notation. This we believe to be an important facilitating factor that notations offer us.

3. SUMMARY

We hope to have introduced some useful ways of thinking about some important aspects of the learning of mathematics that highlight the role of conceptual entities and their relationships with mathematical notations. We regard this chapter as but a beginning into an area of research that others may find productive to pursue in the future.

In §1 we laid out some of the circumstances under which conceptual entities are created and used and what their cognitive function might be, often by pointing to consequences in students' reasoning processes where they have not yet been mentally constructed. We observed three cognitive functions:

- Alleviating working memory or processing load when concepts involve multiple constituent elements, facilitating comprehension of complex concepts,

- the cases of "uniform" operators, "point-wise" operators, and "object-valued operators",

- assisting with the focus of attention on appropriate structure in problem solving.

These functions, undoubtedly, play an important role in mathematical thinking and in fostering the vertical growth of mathematical ideas, at all levels.

In §2 we analyzed the key role that notations play in the entification process by helping substitute names for complex conceptual structures and/or operations. We have discussed three aspects of the interaction between formation of conceptual entities and mathematical notation:

- the role of mathematical notation in forming conceptual entities,

- different types of mathematical notations – *elaborated* and *tacit* notations, and the manner in which they represent conceptual structure,

- notations as substitutes for concepts.

Just as notations can help the formation and application of mental entities, notations can act as *substitutes* for conceptual entities, supplanting the need for them. It is here where both the great power and the great danger in using mathematical notation systems become particularly and unavoidably evident. Accompanying the great power of notations as aids to mathematical thought based on their identity-management role and their structure-substitution role is the great danger that the notations do not refer to any mental content beyond the experienced physical structure of the notations themselves, e.g., as when one deals with an algebraic statement as a character string. This seems to be the case with altogether too many students. While the inventors of notations created them to express and perhaps elaborate their own pre-existing conceptions, in schools we often begin in reverse order, concentrating on manipulation of notations, e.g., the techniques of differentiation and integration in calculus, before providing sufficient experience that would enable the building of mental referents for those notations (Davis, 1986). Students should be given opportunities to build their own notational expressions of their ideas, which can then be guided in the direction of the standard ones. In this way, one builds both notations and conceptions simultaneously, rather than building one or the other first and then attempting to connect the two.

CHAPTER 7

REFLECTIVE ABSTRACTION
IN ADVANCED MATHEMATICAL THINKING

ED DUBINSKY

Our purpose in this chapter is to propose that the concept of reflective abstraction can be a powerful tool in the study of advanced mathematical thinking, that it can provide a theoretical basis that supports and contributes to our understanding of what this thinking is and how we can help students develop the ability to engage in it. To make such a case completely, it would be necessary to do at least several things:

- explain exactly what we mean by reflective abstraction;

- show how it can be used to describe the epistemology of various mathematics concepts;

- indicate how it can suggest explanations of some of the difficulties that students have with many of these concepts;

and

- establish that it can influence the design of instruction in ways that result in a significant improvement in the extent to which students appear to acquire these concepts.

We are certainly not ready to do an exhaustive job on all four of these tasks. Indeed, our main concern here is to make some progress with the first two. There will be a few examples of the third, and we will make reference to other papers in which we have made a start on the fourth especially involving the use of computer activities to help students make mental constructions, with results that are encouraging.

Reflective abstraction is a concept introduced by Piaget to describe the construction of logico-mathematical structures by an individual during the course of cognitive development. Two important observations that Piaget made are first that reflective abstraction has no absolute beginning but is present at the very earliest ages in the coordination of sensori-motor structures (Beth & Piaget, 1966, pp. 203–208)[1] and second, that it continues on up through higher mathematics to the extent that the entire history of the development of mathematics from antiquity to the present day may be considered as an example of the process of reflective abstraction (Piaget, 1985, pp. 149–150).

In the majority of his own work, however, Piaget concentrated on the development of mathematical knowledge at the early ages, rarely going beyond adolescence. What we feel is exciting is that, as he suggested, this same approach can be extended to more advanced topics going into undergraduate mathematics and beyond. It seems that it is possible not

---

[1] Piaget repeated many of his comments on reflective abstraction in several places, but was quite consistent on this topic. Hence, the references we give should be taken as representative.

only to discuss and conjecture, but to provide evidence suggesting, that concepts such as mathematical induction, propositional and predicate calculus, functions as processes and objects, linear independence, topological spaces, duality of vector spaces, duality of topological vector spaces, and even category theory can be analyzed in terms of extensions of the same notions that Piaget used to describe children's construction of concepts such as arithmetic, proportion, and simple measurement.

This is a strong claim embodied in the phrase "can be analyzed" and, before going further, it is necessary to explain what sort of analysis we mean. The goal of our study of reflective abstraction is a general framework which can be used, in principle, to describe any mathematical concept together with its acquisition. We refer to this as a *general theory* of mathematical knowledge and its acquisition. This is the first ingredient of the analysis, but it does not, by itself, lead to any particular description. In addition, the investigator needs to make use of her or his *understanding of the mathematics*. Together these two are enough to obtain a description of any concept but the result would be far too *ex post facto* to expect it to have any relation to how students actually might go about constructing the concept. A third and essential ingredient in the study of any concept is a long drawn-out, time consuming *effort of observation of students* as they try to construct mathematical concepts in order to make sense out of situations in which they find themselves (presumably, but not necessarily, as the result of activities of a teacher). The analysis then consists of a synthesis of these three ingredients brought to bear on the question of how a particular topic in mathematics may be learned. The starting point of our general theory is Piaget's notion of reflective abstraction. Unfortunately, this is not a simple idea clearly explained in one place, but rather something that Piaget appeared to work with over a long period of time after he completed his empirical studies of children in development. It is important, however, that we begin with a solid understanding of what he meant by it before trying to extend it to a wider class of mathematical topics. Therefore we begin this chapter with a section that gives a brief summary of this concept as Piaget elaborated it in a number of books and papers, mostly written in the last 15 years of his life. We will emphasize the construction aspects of reflective abstraction because these are the most important for the development of mathematical thought during adolescence and beyond.

In the second section we will show how Piaget's ideas can be extended and reorganized to form a general theory of mathematical knowledge and its acquisition which is applicable to those mathematical ideas that begin to appear at the post-secondary level and continue to be constructed in the course of mathematical and other scientific research. It is here, in § 2 that we relate various aspects of the general theory to specific topics in advanced mathematical thinking and give several examples of how reflective abstraction can suggest explanations of student difficulties.

Our analysis of a particular mathematical concept leads to what we call a *genetic decomposition* of the concept which is a description, in terms of our theory, and based on empirical data, of the mathematics involved and how a subject might make the constructions that would lead to an understanding of it (which, in our theory, are not very different). It is important to note that we do not suggest that a concept has a unique genetic decomposition or that this is the way every subject will learn it. We only claim that observations of learning in progress form an important source for our genetic decompositions and we offer them as a guide for one possible way of designing instruction. In § 3 we present genetic decompositions for three concepts: mathematical induction, predicate calculus, and function, insofar as we have constructed them. The references given in § 3 contain more information about examples of instructional treatments based on these genetic decompositions, using computer experiences, and about the generally encouraging results of implementing these treatments.

Finally, in § 4 we discuss some of the educational implications of our theory of knowledge and learning and give an overview of how we go about designing an instructional treatment based on it. We feel that the material in this section is very much akin to the ideas in Thompson (1985a).

1. PIAGET'S NOTION OF REFLECTIVE ABSTRACTION

1.1 THE IMPORTANCE OF REFLECTIVE ABSTRACTION

Piaget distinguished three major kinds of abstraction. *Empirical abstraction* derives knowledge from the properties of objects (Beth & Piaget, 1966, pp. 188–189). We interpret this to mean that it has to do with experiences that appear to the subject to be external. The knowledge of these properties is, however, internal and is the result of constructions made internally by the subject. According to Piaget, this kind of abstraction leads to the extraction of common properties of objects and extensional generalizations, that is, the passage from "some" to "all", from the specific to the general (Piaget & Garcia, 1983, p. 299). We might think, for example of the color of an object, or its weight. These properties might be considered to reside entirely in the object but one can only have knowledge of them by doing something (looking at the object in a certain light, hefting it) and different individuals under different conditions might come to different conclusions about these properties.

Pseudo-empirical abstraction is intermediate between empirical and reflective abstraction and teases out properties that the actions of the subject have introduced into objects (Piaget, 1985, pp. 18–19). Consider, for example the observation of a 1-1 correspondence between two sets of objects which the subject has placed in alignment (*ibid*, p. 39). Knowledge of this situation may be considered empirical because it has to do with the objects, but it is their configuration in space and relationships to which this leads that are of concern and these are due to the actions of the subject. Again, of course, understanding that there is a 1-1 relation between these two sets is the result of internal constructions made by the subject.

Finally, *reflective abstraction* is drawn from what Piaget (1980, pp. 89–97) called the *general coordinations* of actions and, as such, its source is the subject and it is completely internal. We will see many instances of reflective abstraction, but a very early example we can mention now is seriation, in which the child performs several individual actions of forming pairs, triples, etc., and then interiorizes and coordinates the actions to form a total ordering (Piaget, 1972, pp. 37–38). This kind of abstraction leads to a very different sort of generalization which is constructive and results in "new syntheses in midst of which particular laws acquire new meaning" (Piaget & Garcia, 1983, p. 299). An example of this is the concept of euclidean ring which is certainly an abstraction and generalization. It might be considered, however, to derive from the properties of a *single* example – the integers.

We can see, therefore, that these different kinds of abstraction are not completely independent. The actions that lead to pseudo-empirical and reflective abstraction are performed on objects whose properties the subject only comes to know through empirical abstraction. On the other hand, empirical abstraction is only made possible through assimilation schemas which were constructed by reflective abstraction (Piaget, 1985, pp. 18–19). Consider, for example a physics experiment which may have the purpose of making an empirical abstraction to obtain factual data about a certain object. The experiment presupposes, however, an enormous range of logico-mathematical preliminaries – in deciding how to pose the question, in the construction of apparatus for "indirect observations" (e.g., triangulation to obtain distances between stars), in the use of particular forms of measurement, and finally, in setting out the results in logico-mathematical language. All of these are concepts that must have been constructed using reflective abstraction. (Piaget, 1980, p. 91). This mutual interdependence can be roughly summarized as follows. Empirical and pseudo-empirical abstraction draws knowledge from objects by performing (or imagining) actions on them. Reflective abstraction interiorizes and coordinates these actions to form new actions and, ultimately new objects (which may no longer be physical but rather mathematical such as a function or a group). Empirical abstraction then extracts data from these new objects through mental actions on them, and so on. This feedback system will be reflected in our extension of these ideas in the next section.

In empirical abstraction the subject observes a number of objects and abstracts a common property. Pseudo-empirical abstraction proceeds in the same way, after actions have been performed on the object. Reflective abstraction, however, is much more complicated. This is not surprising since, according to Piaget, "The development of cognitive structures is due to reflective abstraction..." (Piaget, 1985, p. 143). Before going into the nature of this fundamental process, therefore, we should say a few things about its importance, in Piaget's view, to cognitive thought in general and mathematics in particular.

In two books Piaget (1976, 1978) interpreted the results of many experiments with children in terms of reflective abstraction. But its role is not restricted to the intellectual development of children. From Piaget's psychological viewpoint, new mathematical constructions proceed by reflective abstraction (Beth & Piaget, 1966, p. 205). Indeed, he considered it to be the method by which all logico-mathematical structures are derived (Piaget, 1971, p. 342); and that "it alone supports and animates the immense edifice of logico-mathematical construction" (Piaget, 1980, p. 92).

In support of his position on the role of reflective abstraction in advanced mathematical thinking, Piaget tried to explain a number of major mathematical concepts in terms of the constructions that result from this psychological process. These included the idea of Gödel's incompleteness theorem (Beth & Piaget, 1966, p. 275), the abstract concept of groups (1980, p. 19), Bourbaki's attempts to encompass all of mathematics within three "mother structures" (1970a, p. 24), the general theory of categories (Piaget 1970b, p. 28), the impossibility of constructing the set of all sets (1970b, pp. 70–71), and the mathematical concept of function (Piaget et al, 1977, p. 168). More generally, Piaget considered that it is reflective abstraction in its most advanced form that leads to the kind of mathematical thinking by which form or process is separated from content and that processes themselves are converted, in the mind of the mathematician, to objects of content (Piaget, 1972, pp. 63–64 and pp. 70–71).

Returning to the ideas of Piaget, it is important to emphasize that there is no suggestion here that all (or any) of the advanced mathematics described above is actually done by any kind of direct application (conscious or otherwise) of reflective abstraction. This was not Piaget's purpose in trying to analyze that aspect of thinking. The point, rather, is that when properly understood, reflective abstraction appears as a description of the mechanism of the development of intellectual thought. It is important for Piaget's theory that this same process that describes advanced mathematical thinking appears in cognitive development throughout life from the child's very first coordinations that lead to concepts such as number, measurement, multiplication, and proportion (Piaget, 1972, pp. 70–71). An important ingredient of Piaget's general theory (on which he worked for 60 years) that relates biological evolution to the development of intelligence is the idea that reflective abstraction is one isolated case of certain very general processes that are found throughout living creation (Piaget, 1971, p. 331).

1.2 THE NATURE OF REFLECTIVE ABSTRACTION

As we have seen, reflective abstraction differs from empirical abstraction in that it deals with action as opposed to objects and it differs from pseudo-empirical abstraction in that it is concerned, not so much with the actions themselves, but with the interrelationships among actions, which Piaget (1976, p. 300) called "general coordinations".

According to Piaget, the first part of reflective abstraction consists of drawing properties from mental or physical actions at a particular level of thought (Beth & Piaget, 1966, pp. 188–189). This involves, amongst other things, cognizance or consciousness of the actions (1971, p. 320). It can also include the act of separating a form from its content (1972, pp. 63–64). Whatever is thus "abstracted" is projected onto a higher plane of thought (1985, pp. 29–31) where other actions are present as well as more powerful modes of thought.

It is at this point that the real power of reflective abstraction comes in for, as Piaget observes, one must do more than dissociate properties from those which will be ignored or separate a form from its content (1975a, p. 206). There is "a process which will become increasingly evident over time: the construction of new combinations by a conjunction of abstractions" (Piaget, 1972, p. 23).

Piaget seemed to feel that this construction aspect of reflective abstraction is more important than the abstraction (or extraction) aspect (*ibid*, p. 20). Not only did he assert, as we observed earlier, that construction of this kind is the essence of mathematical development, and that combining formal structures is a natural extension of the development of thought (*ibid*, p. 64), but he also used his analysis of this process to deal with the philosophical question of the nature of mathematical thought (Beth & Piaget, 1966).

Certainly for our purposes, the construction aspect of reflective abstraction will play the major role.

1.3 EXAMPLES OF REFLECTIVE ABSTRACTION
IN CHILDREN'S THINKING

We begin with some of Piaget's examples of reflective abstraction in logico-mathematical thinking at the earlier ages. This is important because of his insistence on the continuity of development as part of his search for a single process or set of processes that related to biological development as well as intellectual development (Piaget, 1971, p. 331). Our suggestion in this chapter is that the specific construction processes that can be used to build sophisticated mathematical structures can be found, already, in the thinking of young children.

commutativity of addition. The discovery that the number of objects in a collection is independent of the order in which the objects are placed requires first that the child count the objects, reorder them, count them again, reorder and count, etc. Each of these actions are interiorized and represented internally in some manner so that the child can reflect on them, compare them, and realize that they all give the same result (Piaget, 1970a, pp. 16–17).

number. According to Piaget (1941), the concept of number is constructed by coordinating the two schemas of classification (construction of a set in which the elements are units, indistinguishable from each other) and seriation (which, as we observed earlier, is itself a coordination of the various actions of pairing, tripling, etc.).

trajectory. The traversal of a path is understood as a coordination of successive displacements to form a continuous whole (Piaget, 1980, p. 90).

see-saw. The balancing of objects on two sides of a see-saw by a combination of actions on *both* sides involves more than just keeping two things in mind at the same time. Because he observed a considerable delay between the time that a child could create the balance and the time that the child appeared to understand how he or she had done it, Piaget saw this as a coordination of two actions into a single system (Piaget, 1978, p. 96).

multiplication. Both psychologically and mathematically, multiplication is the addition of additions. It is, however, objects that are added in the sense that addition is an operation applied to something. In order, therefore, to multiply, it is necessary first to encapsulate the (mental) action of addition into an object (or set of objects) to which addition can be applied (Piaget, 1985, p. 31).

fluid levels. In an experiment asking children to predict the level to which a known amount of fluid would rise in a vessel with sloping sides and markings at equal height divisions (Piaget *et al*, 1977, chapter 7). Piaget pointed out that this situation is a case of "variation of variations". That is, the differential in two vertical markings is a variation, but the amount of change also varies because of the sloping sides. Hence, the first variation must become an object to which an action is applied (sloping sides) resulting in a higher order variation.

1.4 VARIOUS KINDS OF CONSTRUCTION
IN REFLECTIVE ABSTRACTION

In considering the above examples of reflective abstraction as methods of construction, we can isolate four different kinds which will be important for advanced mathematical thinking. We add a fifth which Piaget considers at length, but was not, for him, part of reflective abstraction.

- With the appearance of the ability to use symbols, language, pictures, and mental images, the child performs reflective abstractions to represent (Piaget, 1970a, p. 64), that is, to construct internal processes as a way of making sense out of perceived phenomena. Piaget called this *interiorization* (1980, p. 90) and referred to it as "translating a succession of material actions into a system of interiorized operations" (Beth & Piaget, 1966, p. 206). The commutativity of addition described above is one example of this. (See also Thompson, 1985a, p. 197.)

- Several of our examples such as trajectory and see-saw involve the composition or *coordination* of two or more processes to construct a new one. This is to be distinguished from Piaget's phrase, "general coordinations of actions" which refers to all ways of using one or more actions to construct new actions or objects.

- Multiplication, proportion and variation of variation exemplify the construction which is perhaps the most important (for mathematics) and most difficult (for students). This is *encapsulation* or conversion of a (dynamic) process into a (static) object. As Piaget put it (1985, p. 49), "... actions or operations become thematized objects of thought or assimilation". He considered that "The whole of mathematics may therefore be thought of in terms of the construction of structures, ... mathematical entities move from one level to another; an operation on such 'entities' becomes in its turn an object of the theory, and this process is repeated until we reach structures that are alternately structuring or being structured by 'stronger' structures" (Piaget, 1972, p. 70). From a philosophical point of view, Piaget was applying the idea of encapsulation to the relativity between form and content when he referred to "...building new forms that bear on previous forms and include them as contents" and "reflective abstractions that draw from more elementary forms the elements used to construct new forms" (Piaget, 1985, p. 140).

- When a subject learns to apply an existing schema to a wider collection of phenomena, then we say that the schema has been *generalized*. This can occur because the subject becomes aware of the wider applicability of the schema. It can also happen when a process is encapsulated to an object as, for example, the ratio of two quantities, or addition, so that an existing schema such as equality or addition can then be applied to it to obtain, respectively, proportion or multiplication. The schema remains the same except that it now has a wider applicability. The object changes for the subject in that he or she now understands that it can be assimilated by the extended schema. Piaget referred to all of this as a reproductive or generalizing assimilation (1972, p. 23), and he called the generalization *extensional* (Piaget & Garcia, 1983, p. 299).

Once a process exists internally, it is possible for the subject to think of it in reverse, not necessarily in the sense of undoing it, but as a means of constructing a new process which consists of *reversing* the original process. Piaget did not discuss this in the context of reflective abstraction, but rather in terms of the INRC group. We include it as an additional form of construction.

2. A THEORY OF THE DEVELOPMENT OF CONCEPTS IN ADVANCED MATHEMATICAL THINKING

2.1 OBJECTS, PROCESSES, AND SCHEMAS

Although, as we have pointed out, Piaget believed that reflective abstraction was as important for higher mathematics as it was for children's logical thinking, his research was mainly concerned with the latter. In order to try to develop the notion of reflective abstraction for advanced mathematical thinking, we will isolate what seem to be the essential features of reflective abstraction, reflect on their role in higher mathematics, and reorganize or reconstruct them to form a coherent theory of mathematical knowledge and its construction.

For us, reflective abstraction will be the construction of mental objects and of mental actions on these objects. In order to elaborate our theory and relate it to specific concepts in mathematics, we will use the notion of *schema*. A schema is a more or less coherent collection of objects and processes. A subject's tendency to invoke a schema in order to understand, deal with, organize, or make sense out of a perceived problem situation is her or his knowledge of an individual concept in mathematics. Thus an individual will have a vast array of schemas. There will be schemas for situations involving number, arithmetic, set formation, function, proposition, quantification, proof by induction, and so on throughout all of the subject's mathematical knowledge. Obviously, these schemas must be interrelated in a large, complex organization. For example, there will be a proof schema, which can include a schema for proof by induction. This latter in turn could include a schema for proposition valued functions of the positive integers (see p. 112). Hence there would be a relation with the schemas for number, for function, and for proposition. On the other hand, there is a sense in which a proof is an action applied to a proposition, so that the proof schema might be one of the processes in the proposition schema.

We will also sometimes use the term *process* or *mental process* instead of mental action when we are emphasizing its internal (to the subject) nature. Finally the term *object* will refer to a mental or physical object (avoiding any discussion of the nature of the distinction).

One of our goals in elaborating the general theory is to isolate small portions of this complex structure and give explicit descriptions of possible relations between schemas. When this is done for a particular concept, we call it a *genetic decomposition* of the concept. We should also point out that although we only give, for each concept, a single genetic decomposition, we are not claiming that this is *the* genetic decomposition, valid for all students. Rather it represents one reasonable way that students might use to construct a concept.

It is not easy to separate a description of mathematical knowledge from its construction. As Piaget put it, "... the problem of knowledge, the so-called epistemological problem, cannot be considered separately from the problem of the development of intelligence" (Piaget, 1975a, p. 166). It is not possible to observe directly any of a subject's schemas or their objects and processes. We can only infer them from our observations of individuals who may or may not bring them to bear on problems – situations in which the subject is seeking a solution or trying to understand a phenomenon. But these very acts of recognizing and solving problems, of asking new questions and creating new problems are the means (in our opinion, essentially the *only* means) by which a subject constructs new mathematical knowledge.

This is where reflective abstraction comes in. Thus, although we might say that mathematical knowledge consists of a collection of schemas, we have little to say about how that knowledge exists inside a person. It does not seem to reside in memory or in a physiological configuration. All we can say is that a subject will have a propensity for responding to certain kinds of problems in a relatively (but far from totally) consistent way which we can (as far as our theory has been developed) describe in terms of schemas. When the subject is successful, we say that the problem has been assimilated by the schema. When the subject is not successful then, in favorable conditions, her or his existing schemas may be *accommodated* to handle the new phenomenon. This is the constructive aspect of reflective abstraction to which we referred as forming the main object of our concern.

In this sense, the study of reflective abstraction is complementary to investigation of notions such as *epistemological obstacles* as studied by Cornu (1983), and Sierpińska (1985a, 1985b) or the conflict between *concept image* and *concept definition* as investigated by Schwarzenberger & Tall, 1978; Tall & Vinner, 1981; Dreyfus & Vinner, 1982; Vinner, 1983; Tall, 1986a; Vinner & Dreyfus, 1989). One can think of reflective abstraction as trying to tell us what needs to happen whereas the other notions attempt to explain why it does not. It is possible that our idea of using computer experiences (Ayers *et al*, 1988; Dubinsky, 1990a, 1990b) to help students make reflective abstractions can be a way of dealing with these obstacles and conflicts. But these are matters for other investigators and other papers.

2.2 CONSTRUCTIONS IN ADVANCED MATHEMATICAL CONCEPTS

In the previous section, we isolated five kinds of construction that Piaget found in the development of children's logical thinking: interiorization, coordination, encapsulation, generalization, and reversal. We will reconsider each of them in the context of advanced mathematical thinking to describe how new objects, processes and schemas can be constructed out of existing ones.

Some of the following examples will apply to a single one of the five kinds of construction and others will apply to a combination of two or more of them. Some of the statements we make are based on observations of students and others are only suppositions, derived as a preliminary to observations, from the general theory and our knowledge of the mathematics.

As we make these statements about constructions that we have seen students appear to make or that our investigations suggest they need to make, or as we conjecture that certain concepts could be constructed in these ways, the reader should be aware that we are not suggesting that it is automatic, natural, or easy for students to take these steps. An important aspect of the whole problem of education that we do not consider in this paper is to explain why students do or do not make these particular constructions and what can be done to help them. This is an important issue for research in mathematics education.

An important part of understanding a function that we have observed is to construct a process (Dubinsky *et al*, 1989). For individual examples this means that the subject responds to a situation in which a function may appear (via formula, as an algorithm, or represented by data) and for which there is a process by which the value of the function, for a particular value in the domain, is obtained. Given such a situation, the subject may respond by constructing, in her or his mind, a mental process relating to the function's process. This is a prime example of interiorization.

An example of the same kind of mental activity in a completely different mathematical situation could arise in understanding proofs. When the mathematician exclaims (as which of us has not?) that "I can understand each step of the proof, but I don't see the whole picture", it could be the case that he or she is expressing the necessity of interiorizing a whole collection of processes and coordinating them to obtain a single process. The interiorization of the total process can be, in our opinion, the final step in "making a proof your own".

Interiorization may not always be difficult. Most students seem to have little trouble with constructing a mental process for multiplying a matrix and a vector, or two matrices. This could be because there is a straightforward "hand-waving" action, used by most teachers, that is a physical representation of the multiplication and could form an intermediary between the external action and its interiorization. It seems that mathematics becomes difficult for students when it concerns topics for which there do not exist simple physical or visual representations. One way in which the use of computers can be helpful is to provide concrete representations for many important mathematical objects and processes (see Chapter 14).

Turning now to coordination, one of the most important examples that we have seen occurs in the formation of the composition of two functions. Based on our research (Ayers *et al*, 1987; Dubinsky *et al*, 1989), we would like to propose the following psychological description. Composition is a binary operation which means that it acts on two objects to form a third. Thus, it is necessary to begin with two functions, considered as objects. The subject must "unpack" these objects, reflect on the corresponding processes, and interiorize them. Then the two processes can be coordinated to form a new process that can then be encapsulated into an object which is the function that results from the composition. This is much more complicated than simple substitution and perhaps explains why students have so much difficulty with ideas like the chain rule for differentiation, in which it would be necessary to coordinate this view of composition with the notion of derivative. It could also explain those results of Ayers *et al* (1988) in which students seem to improve their understanding of composition as a result of performing computer tasks designed to foster these mental operations.

A whole class of examples of that could be described as coordination of schemas in advanced mathematics is given by the "mixed" structures: topological vector spaces, differentiable manifolds, homotopy groups, etc.

It is often possible to observe students having difficulty with determining the cardinality of sets such as

$$\{4, \{-3, 2, -1/7\}, \{\{17, 5\}\}\}.$$

Very often, even undergraduates will think that this set has 6 elements (rather than 3). We suggest that the difficulty is that the students have not encapsulated the sets $\{-3, 2, -1/7\}$, $\{17, 5\}$ into objects so as to understand the nested structure of the given set.

The indefinite integral forms an important example that can be interpreted as encapsulation together with interiorization. Estimating the area under a curve with sums and passing to a limit is, of course, a process. Students who seem to understand this often have difficulty with the next step of varying, say, the upper limit of the integral to obtain a function. What is lacking, we suggest, is the encapsulation of the entire area process into an object which could then vary as one of its parameters vary. This would then form a "higher-level" process which specifies the function given by the indefinite integral. The complexity of this total process might then explain why students have such difficulty with not only the Fundamental Theorem of Calculus, but such powerful definitions as

$$\log(x) = \int_1^x \frac{dt}{t}, \ x > 0$$

A rather pervasive example that can be interpreted as encapsulation in mathematics is duality. The dual of a vector space, for example, is obtained by considering all of the linear transformations from the space to its scalar field as objects, collecting them in a set, and introducing a natural algebraic structure on this set. It seems to us that this is an act of encapsulation that is essential in this branch of mathematics.

The simplest and most familiar form of reflective abstraction is generalization. According to our investigations, we can say that a subject's function schema, in which functions transform numbers, is generalized to include functions which transform other kinds of objects (once they have been encapsulated) such as vectors, sets, propositions, or other functions. Similarly it would seem that the schema of factorization of positive integers can be generalized in this way to factoring polynomials, and then to an arbitrary euclidean ring. Vectors of dimension two and three can be generalized to include higher, and even infinite, dimensional vectors. All of these and a host of other examples in mathematics seem to involve the application of an existing schema, essentially unchanged, to new objects (which are often the result of encapsulation).

Finally there is reversal of a process. We can mention a number of familiar activities in mathematics that appear to involve the reversal of a process: subtraction and division, solving an equation, inverting a function, proving an inequality (in which one often starts with the conclusion, manipulates until something known to be true is obtained, and then sees if the argument can be reversed), and the mysterious choice of expressions such as $\frac{\varepsilon}{2\sqrt{M}}$ in proving limit theorems.

2.3 THE ORGANIZATION OF SCHEMAS

In the previous section, we suggested how the construction of various concepts in advanced mathematics could be described in terms of the five forms of construction in reflective abstraction: interiorization, coordination, encapsulation, generalization, and reversal. We offer the conjecture that the construction of all mathematical concepts can be described in these terms. It may be that additional forms of reflective abstraction will have to be added as additional concepts are considered, but we suggest that the five given here tell something like the full story.

Of those concepts (mathematical induction and predicate calculus) for which we have made a more or less complete genetic decomposition (Dubinsky, 1986; Dubinsky, Elterman & Gong, 1988), our analysis has been greatly influenced by data obtained from observations (interviews, written tasks, computer work, etc.) of students while they are trying to understand the concept in question. The genetic decomposition is then derived from a synthesis of these empirical results, our general theory, and our mathematical knowledge of the concept in question. This is why it takes a long time and has only been done extensively for two concepts. Work on other concepts (e.g., function, limit) is proceeding slowly and, we hope, deliberately.

The following description of the organization of a schema is just a summary of what we have seen in the concepts investigated thus far and, therefore, is somewhat tentative. We give it here in general terms and then, in the next section, see how it looks in the context of mathematical induction and predicate calculus. In addition, with more anticipation than certainty, we will suggest how it might look for the concept of function, after considerably more data has been gathered.

The structure of a schema is displayed in figure 13.

As we have already indicated, one should not think of a schema statically, but rather as a dynamic activity (or propensity for such activity) by the subject. Moreover, the existence of a schema is inseparable from its continuous construction and reconstruction. Thus, in describing the system in Figure 1 we will try to do several things simultaneously: describe what is there, describe what happens, describe how things are constructed, and refer to some of the examples we have discussed previously. An additional complication is that, as indicated in the picture, a schema is not a linear list of items but rather a circular feedback system. Our description, necessarily linear, must break in at some point. In any case, the following discussion is an alternative way of organizing the five kinds of construction analyzed in the previous two sections. Here we also include the results of the constructions (objects and processes).

We begin with *objects*. These encompass the full range of mathematical objects: numbers, variables, functions, topological spaces, topologies, groups, vectors, vector spaces, etc., each of which must be constructed by an individual at some point in her or his mathematical development.

At any point in time there are a number of *actions* that a subject can use for calculating with these objects. These actions go far beyond numerical calculation resulting in numerical answers. The computation of the homotopy group of a topological space is a calculation. So is the determination of the (topological) dual of a (locally convex topological) vector space. We will return to this example a few paragraphs below when we discuss coordination.

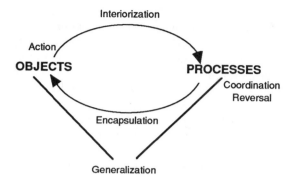

Figure 13: Schemas and their construction

It is possible for a subject to work with actions in ways other than just applying them to objects. First, an action must be interiorized. As we have said, this means that some internal construction is made relating to the action. An interiorized action is a *process*. Interiorization permits one to be conscious of an action, to reflect on it and to combine it with other actions. For example, the computation of the dual of a particular vector space is an action on that object. The idea, independent of any particular vector space, that it may have a dual and it can often be computed, is the process that results from interiorizing this action.

Interiorizing actions is one way of constructing processes. Another way is to work with existing processes to form new ones. This can be done, for example, by reversal. A calculus student may have interiorized the action of taking the derivative of a function and may be able to do this successfully with a large number of examples, using various techniques that are often taught and occasionally learned in calculus courses. If the process is interiorized, the student might be able to reverse it to solve problems in which a function is given and it is desired to find a function whose derivative is the original function. This is anti-differentiation or integration, and it too, is first an action and then must be interiorized to become a process. Encapsulating both the differentiation and integration processes – at least to the point of having them as objects of reflection – would seem to be an essential prerequisite for understanding the fundamental theorem of calculus.

Another way of making new processes out of old ones is to compose or coordinate two or more processes. For example, let us return to the dual of an infinite dimensional vector space and imagine (this is purely conjectural) how a subject might think about it. A subject may have a schema (discussed in the previous section) for constructing the dual of a finite dimensional vector space. If an *infinite* dimensional vector space comes along, then it seems that exactly the same schema can be used to construct its dual, as well. We would say that the new phenomenon (infinite dimensional vector space) has been assimilated to this schema. As mathematical experience goes further, however, this result would not be very satisfactory, and it is particularly convenient to make use of topological structures. If there is, in the subject's schemas, a process for equipping a set with a topology, then this could be coordinated with the vector space schema to obtain a topological vector space. Now within a schema for topological space there should be a schema for the concept of continuous function and within a vector space schema there should be a notion of linear function. Coordinating continuity and linearity, one can obtain the idea of a continuous linear function. This coordination would permit the subject to extend and reorganize the process for constructing the dual of a vector space to apply to the set of those functions from the original set to the scalar field which are continuous as well as linear, thereby obtaining the topological dual. In such a situation we would say that the schema for duals has been accommodated to the new phenomena (involving topologies) and experiences which made the old schema less than satisfactory.

In addition to using processes to construct new processes, it is also possible to reflect on a process and convert it into an object. Anytime a *set* of functions is considered, it seems necessary to think of the functions in question as objects. Initially, functions are processes and so the subject must have performed an encapsulation in order to consider them as objects. It is important, for example in composition of functions, for the subject to alternate between thinking about the same mathematical entity as a process and as an object. (*cf.* p. 104.)

A more advanced, and yet more fundamental example where encapsulation may occur is in the concept of a topology. Initially, there is the notion of nearness or convergence, which is a process. One of the accomplishments of twentieth century mathematics is to capture this idea with the device of a collection of subsets (so-called "open sets") which must satisfy certain conditions but is otherwise arbitrary. The interaction (really another form of coordination) between, on the one hand, a collection of sets which may be taken as arbitrary in order to investigate general topological properties related to but not identical with notions of "nearness", and on the other hand, a very specific choice of this collection so as to apply those properties to important concrete situations, say in analysis, and the use of the resulting observations to stimulate the development of further general properties, and so on, has led to a great deal of important new mathematics of both abstract and concrete natures. A key step in this progress may be described as the encapsulation of the process "nearness" to the object "topology".

We conclude this section with a recapitulation of our description of the construction of schemas in the context of the example, already mentioned on several occasions above, of the (topological) dual of a (topological) vector space. This suggestion of a genetic decomposition for the concept of dual is totally speculative in the sense that it depends entirely on our theory and our understanding of the relevant mathematics. We have gathered no data (other than introspection on our own experience) to support our suggestions. On the other hand, it may be interesting for those with a background in mathematics to see that our theory at least *appears* to be reasonably compatible with a topic from the arena of mathematical research. It is an important point that the same ideas that described the thinking of young children and adolescents can be used to talk about higher mathematics.

In the beginning, there are vectors, which are the objects, and actions on vectors including addition, scalar multiplication and the gathering together of vectors in a set with these operations, to form a *vector space*. This is a schema that we assume the subject possesses. We also assume that the subject has a schema for functions that transform numbers into other numbers.

The first step, according to our conjecture, is to generalize the function schema to include as a function any process that transforms vectors into scalars. This could then be coordinated with the addition of vectors and their multiplication by scalars to restrict the functions to processes that transform vectors into scalars, but preserve the algebraic operations of addition and scalar multiplication.

We would then say that these functions are encapsulated into objects called *linear functionals* and collected together in a single set. At this point we would like to suggest that, although the assigning of a name like linear functional to a process is closely connected with its encapsulation into an object, it is the encapsulation that is fundamental and gives "meaning" to the name. To name processes without encapsulating them is the essence of jargon. When there is a complaint that a particular discourse has too much terminology and not enough meaning, we feel that the real difficulty is that labels are being assigned without an opportunity for encapsulating that which is being labeled.

In any case, the set of linear functionals can be assimilated to the vector space schema (which may have to be accommodated to this purpose – that is, it may be necessary to project and reconstruct this schema on the higher level of a vector space whose elements are linear functionals) by defining addition and scalar multiplication of these functionals. This can be done very naturally, interpreting the functions as processes and using "point-wise operations". In this way, the set of linear functionals becomes a vector space, called the algebraic dual.

Now comes a major interiorization. What we have been describing is an action applied to a vector space E that constructs its algebraic dual E^*. When this has been interiorized, one has constructed the beginning of duality theory. One can reverse the process to look for a "pre-dual", that is, given a vector space F, can one find a vector space E whose algebraic dual is F? (The answer is yes if E is "finite-dimensional", but otherwise it may or may not be possible.) Or one can perform the process twice. When two instantiations are coordinated, one obtains the bidual E^{**}. The concept of reflexivity (fairly simple in the case of the algebraic dual) has to do with whether $E=E^{**}$.

Next, as we mentioned above, topology and algebra can be coordinated to obtain the concept of topological vector space and the schema for dual can be projected onto this higher plane and reconstructed by introducing considerations of continuity, to obtain the topological dual E' of a topological vector space E.

Again the action of constructing the topological dual can be interiorized into a process and the concepts of pre-dual and reflexivity (much more interesting in the topological case) can be reconstructed and their properties investigated. Even more interesting, the *content* of forming the topological dual can be removed from the *form* of this process (by reflecting on it) and this would give rise to the idea of dual pairs $\langle E,F \rangle$ in which algebra and topology are mixed in free and varying combinations to obtain the modern theory of dual systems in linear topological spaces.

3. GENETIC DECOMPOSITIONS OF THREE SCHEMAS

We will consider three schemas in some detail: mathematical induction, predicate calculus, and function. Our goal is to show how the general theory elaborated in the previous section can be used in possible descriptions of the nature and construction of these specific schemas. Thus in each case we will point out the relevant objects and processes as well as the instances of reflective abstractions that seem to us can be used in constructing them.

The details that we are about to present come from our three sources. First, there is the psychological data that we have gathered through observations of students in the midst of trying to learn these concepts. These experiments are described in full detail in Dubinsky (1986, in press a,b), Dubinsky *et al* (1986, 1989, in press). This data, along with the ideas of Piaget formed the basis for the derivation of our theory, which is the second source of the genetic decompositions. That is, for each phenomenon that was observed, we tried to use our theory to describe it, adjusting the theory when necessary. (As the necessity for adjustment occurs less often, our confidence in the theory increases.) The third source of the descriptions is our mathematical understanding of the concepts in question. It seems important that a genetic decomposition should make sense from a mathematical point of view, although it might not be exactly how the mathematician might have analyzed the subject in thinking about how to teach it.

These three sources actually only apply in full to the first two examples: mathematical induction and predicate calculus. Because our data, and the analysis that leads to our conclusions, already appears in the above references, we do not repeat it here. In the case of function, we have begun to gather data, but our studies were not yet complete at the time of writing and so we make some mention of it, although very limited. Thus the genetic decomposition of function given here is based mainly on the theory and our mathematical understanding of function. As such, it must be taken as speculation that may form a bridge for future work. As we obtain and analyze data on students' learning the concept of function, it will be interesting to see how close the genetic decomposition postulated here comes to what is derived when the genetic data are taken into account. In a sense, this can provide an indication of the predictive value of our theory as it has been developed so far.

3.1 MATHEMATICAL INDUCTION

The aspect of induction that we are interested in has to do with a subject's understanding of the induction process, why it "works" to establish something and how to construct an induction proof. Ultimately, this has to be coordinated with a notion of infinity but it may be that understanding the induction process is a precursor to constructing a notion of infinity. It would be an interesting investigation to apply, to the concept of infinity, our method of helping students learn induction (Dubinsky, 1986, in press).

In the first instance, mathematical induction is a process in that one interiorizes the actions of moving along (as "n increases") from one proposition to the next and, after an initial independent determination, establish the truth of a statement by applying a tool (truth of an implication) that was previously constructed.

Mathematical induction is also an object in the subject's general schema for proofs. This means that the induction process must have been encapsulated in order that the subject can reflect on it, along with other methods, when confronted with a theorem to prove, so as to select induction as the method for a particular problem.

The method itself is constructed by working with two major schemas: function and logic. The developments of these two schemas are intertwined through various coordinations. We can illustrate the process with a chart as shown in figure 14.

We start with the assumption that the subject possesses a function schema and a logic schema that are already developed to the point where, for example, the function schema

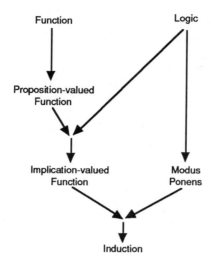

Figure 14: Genetic Decomposition of Mathematical Induction

includes the ability to construct a process relating to a particular transformation of numbers (see § 3.3), and the logic schema can construct statements in the first order propositional calculus (see § 3.2). In particular we assume that the function schema includes the process of evaluation of a function for a given value in its domain and that the logic schema includes a process for logical necessity, that is, in certain situations, the subject will understand that if A is true then of necessity B will be true. Of course we are not asserting that the subject will necessarily be aware of these schemas in this terminology. What we mean, for example, is that the subject will be able to think in terms of plugging a value of a positive integer into a statement and asking if the result is a true statement. This is a function and we can infer from a subject's actions that it may exist in her or his mind as a schema – but we would hardly require young subjects to be aware of it as such in order to understand induction.

The formation of first order propositions is a process in the logic schema which can come from interiorizing actions (conjunctions, disjunctions, implications, negations) on declarative statements (objects). The subject can perform a reflective abstraction on this process to obtain new objects which are the propositions of the first order propositional calculus, on which the same actions can be performed. Consider for example, a simple proposition such as,

$(P \vee Q) \wedge R$

where P, Q and R are simple declarations. The formation of the disjunction $P \vee Q$ can be described as an action on the statements P, Q. It is not just the action of putting these symbols in this expression. The subject must also construct a mental image involving the two statements and the determination of the truth or falsity of the disjunction in various situations. If nothing further is done after this action is interiorized, then it will not be possible to combine this with R to get the full proposition. First, the disjunction process must be encapsulated to form a new object $(P \lor Q)$ which is a statement that can be conjoined with another statement, such as R. Note how the use of parentheses in mathematical notation corresponds to encapsulation.

Iterating this procedure, the subject enriches her or his logic schema to obtain a host of new objects consisting of first order propositions of arbitrary complexity. Next the function schema comes in. We are assuming that this schema can be used by the subject to construct processes that transform numbers (for example an integer) into other numbers. It must be generalized to permit the subject to construct processes that transform positive integers into propositions, to obtain what we shall call a *proposition valued function of the positive integers*. Consider for example, a statement such as,

Given a number of dollars, it is possible to represent it with $3 chips and $5 chips.

For such a statement, the subject must construct a process whereby, for each positive integer n, a proposition is constructed which is the same statement, but with "a number of dollars" replaced by that value of n. This is the proposition valued function. In order to evaluate it, the subject must construct another process whereby, given n, a search is made and it is determined whether it is possible to find non-negative integers k, j such that

$n = 3j + 5k.$

It is useful for the subject to discover that the *value of this function* is *true* for $n = 3, 5$, *false* for $n = 1, 2, 4, 6, 7$ and then appears to be *true* for all higher values.

It is only at this point that the subject can realize that the problem of "proving" the statement consists of determining that the value of the function is *true* for *all* values of $n \geq 8$. For this, the proof schema can be invoked. If it contains the schema for induction, it can be used, if not, further (re-)construction must take place. In describing this construction, we reiterate that, in the context of this theory, it is never clear (nor can it be) whether one is talking about a schema that is present or one that is being (re-)constructed.

Before going on with the description, there is a side issue that should be considered. Whether the subject is able to construct a proposition valued function of the positive integers to deal with a particular statement depends not only on the existence of the schemas we are talking about, but also may require additional knowledge about the particular situation – so-called "domain knowledge". Thus, although the above example of chips is probably well within the domain knowledge of most students who find themselves trying to learn induction, others may not be. We have found, for example, that the following statement provides difficulty for university undergraduates.

An integer consisting of 3^n identical digits is divisible by 3^n.

The trouble could lie in understanding the relationship between the value of an integer and its representation with digits. It is a sort of "grown-up" version of the difficulty with the concept of place value and it suggests that many students have not really constructed this concept – at least in a sufficiently powerful form.

Returning now to the construction of proof by induction, the next development provides an example of a cognitive step which our research has pointed out as providing a serious difficulty, whereas if one takes only the mathematical point of view, there is not even a step that needs to be taken. This is the case even though it relates to an overt difficulty encountered by everyone who has tried to teach mathematical induction.

We are referring to the notion that the essential point in an induction proof is that one does not prove the original statement directly, but rather the implication between two statements derived from it. This is the major difficulty for students. It requires a cognitive step which is not necessary as a mathematical step. To explain, let us denote by P the proposition valued function to be proved. Now $P(n)$ can be any proposition, in particular, it can be an implication. Therefore, if we define the proposition valued function Q by

$$Q(n) = (P(n) \Rightarrow P(n+1))$$

then, *from a mathematical point of view* there is nothing new in Q, that is, once one understands P then, as a special case, one understands Q. We have observed, however, that with students, this is not the case from the *cognitive point of view*. In the first place, implications are the most difficult propositions for students and they are generally the last to be encapsulated. Furthermore, there is a difference between constructing P from a given statement and constructing Q *from P*. This is the step which must be taken. If there is some subtlety here, then it might help explain the difficulty that students have at precisely this point.

To summarize, this step appears to require the encapsulation of the process of implication so that an implication is an object and can be in the range of a function, the generalization of the function schema to include implication valued functions, and the interiorization of the process of going from a proposition valued function of the positive integers to its corresponding implication valued function.

The next step is to add to the logic schema a process which we shall call *modus ponens*. This process is an interiorization of an action applied to implications (assuming as above that they have been encapsulated into objects). The action consists of beginning at the hypothesis, determining that it is true, and then "crossing the bridge" to the conclusion and asserting its truth.

Finally, there is a coordination of the function schema, as it applies to an implication valued function Q (obtained from a proposition valued function P) and the logic schema as it applies to the process modus ponens which has just been constructed. Included in the function schema is the process of evaluation, that is, sampling values n of the domain (positive integers in this case) and computing the value of the function, $Q(n)$, that is, $P(n) \Rightarrow P(n+1)$. Suppose that it has been established that Q has the constant value true. The first step in this new process which must be constructed is to evaluate P at 1 and to determine that $P(1)$ is true (or, more generally, to find a value n_0 such that $P(n_0)$ is true). Next, the function Q is evaluated at 1 to obtain $P(1) \Rightarrow P(2)$. Applying modus ponens and the fact (just established) that $P(1)$ is true yields the assertion $P(2)$. The evaluation process is again applied to Q but this time with $n=2$ to obtain $P(2) \Rightarrow P(3)$. Modus ponens again gives the assertion $P(3)$. This is repeated ad infinitum, alternating the processes of modus ponens and evaluation. Thus we have a rather complex coordination of two processes that we believe leads to an *infinite* process.

This infinite process is encapsulated and added to the proof schema as a new object, proof by induction.

3.2 PREDICATE CALCULUS

The predicate calculus schema appears to be obtained through a reconstruction of a schema resulting from coordinating a schema for first order propositional calculus with a function schema. The construction is illustrated in Figure 15. According to this analysis, the objects in the propositional calculus schema are the propositions. The most important process is the determination of the truth or falsity of a proposition. Other processes include the formation of new propositions by the standard logical operations such as conjunction, disjunction, implication and negation. They also include the process of expressing an English statement in the formal language of symbolic logic and translating from that syntax back to English. Then of course there are all the usual tasks that students are asked to perform such as manipulation of the formulas, construction of truth tables, determination of the validity of arguments and so on. Finally, we can mention the process of reasoning about a statement, for example, to know if the truth or falsity of the statement

$$(P \Rightarrow Q) \vee (\text{not } (Q \wedge R))$$

is determined once you know that $P \Rightarrow R$ is false.

Amongst the various manipulations of logical expressions, one in particular will be important in the sequel. That is the process of applying the conjunction operation ("and" or $\wedge$) to a set of propositions as in

$$(x_1 > b_1) \wedge (x_2 > b_2) \wedge \ldots \wedge (x_n > b_n).$$

There is a similar process for disjunction ("or" or $\vee$). This is a manipulation of symbols, but there is an underlying process connected with the truth value of the resulting proposition.

In a sense, the objects in the first order propositional calculus are constants. In an expression such as $(P \Rightarrow Q) \vee (\text{not } (Q \wedge R))$ the quantities P, Q and R are constants whose value may be unknown, but fixed. The subject's thinking about such matters can be elevated to a higher plane when the propositional calculus schema is coordinated with the function schema (appropriately reconstructed on this higher plane) to consider such an expression as determining a function — in this case of the three variables, P, Q and R. This is the beginning of the predicate calculus schema. Of course, a part of this coordination and reconstruction was discussed already in the previous section for the special case of proposition valued function of the positive integers.

As before, an important new process that can be constructed is the iteration (in the subject's mind) through the domain of a proposition valued function, checking the truth or falsity of the resulting proposition for each value of the variable. Consider, for example a statement such as

Given a car in the parking lot, if the tire fits the car, then the car is red.

Here, *tire* may be considered to be a constant, but *car* should be thought of as a variable whose domain is the set of cars in the parking lot. There is an obvious action of walking through the parking lot, checking each car to see if the tire fits and, if it does, seeing if the car is red. When such a statement appears in a mathematical context, as in

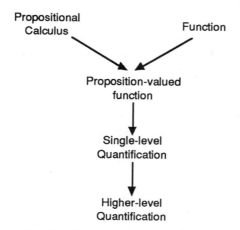

Figure 15: Genetic Decomposition of Predicate Calculus

Given $x \in$ domain(F), if $|x-x_0| \le \delta$, then $|F(x)-F(x_0)| \le \varepsilon$

then the mental process seems to consist in looking at each $x \in$ domain(F) to see if $|x-x_0| \le \delta$ and, if so, seeing if $|F(x)-F(x_0)| \le \varepsilon$.

This iteration process must now be coordinated with one of the two processes we mentioned earlier: applying conjunction or disjunction to a set of propositions. The resulting process can be encapsulated which leads to a single existential or universal quantification as in

For all cars in the parking lot, if the tire fits the car, then the car is red.

$\forall\, x \in$ domain(F), $|x-x_0| \le \delta \Rightarrow |F(x)-F(x_0)| \le \varepsilon$

We call this a *single-level quantification.*

The single-level quantification creates new objects which are again propositions so that all of the previous processes of logical operations, negation and reasoning about statements are reconstructed on this higher plane. Particularly important for understanding many mathematics topics is the interiorization of a statement given as a quantification. The subject seems to need a strong mental image of the iteration and quantification process that we have described in order to relate the statement to the mathematical situation that is being considered.

Next comes *two-level quantifications* in which two (usually different type) quantifiers are applied in succession to a proposition valued function of two variables. For example, the statements we have considered may be extended to obtain,

There is a tire in the library such that for all cars in the parking lot, if the tire fits the car, then the car is red.

or

$\exists \delta > 0 \ni \forall\, x \in$ domain(F), $|x-x_0| \le \delta \Rightarrow |F(x)-F(x_0)| \le \varepsilon$.

The process which we just described for constructing single-level quantifications ended with an encapsulation so that the result becomes a proposition which is a mental object. Note that the effect of a quantification is to eliminate a variable. If the original proposition valued function had two variables, then the resulting object actually depends on the value of the other variable and the schema for single-level quantifications can again be applied to this proposition valued function. For example, in the case of the tires and cars, the universal quantification over cars results in a proposition valued function of the single variable, *tire*. This function can then be subjected to an existential quantification to obtain a single, constant proposition. Thus, when analyzing a statement which requires a two-level quantification over two variables, the subject can begin by parsing it into two quantifications. There is an inner quantification over one of the variables in a proposition valued function of two variables. There is also an outer quantification over the other variable. What we have described is a coordination of these two quantifications to obtain a process which will be a two-level quantification. In order to proceed to higher-level quantifications this new process is again encapsulated to obtain a new object. Once it is encapsulated, it can then be subjected to the same processes (thereby generalized) as were the single level quantifications.

Given a statement which is a three-level quantification, such as the definition of continuity of F at x_0,

$$\forall \varepsilon > 0, \exists \delta > 0 \ni \forall x \in \text{domain}(F), \ |x-x_0| \le \delta \Rightarrow |F(x)-F(x_0)| \le \varepsilon.$$

the subject can group the two inner quantifications and apply the two-level schema to again obtain a proposition which depends on the outermost variable (in this case ε). This proposition valued function is then quantified as before to obtain a single proposition. The entire procedure can now be repeated indefinitely to obtain quantifications of any level. At each level, the same processes of logical operations, negation, reasoning, etc. are reconstructed.

3.3 FUNCTION

As we indicated earlier, the thoughts about the function concept given here are based mainly on the general theory and our understanding of this concept from the mathematical point of view. Our purpose for including it and giving some examples of preliminary data is to illustrate the explanatory power of our theory and to set guideposts for subsequent empirical work. In the past decade, the function concept has been investigated by a number of authors in ways that are quite different from the approach described here (see especially Dreyfus & Eisenberg, 1983, 1984; Dreyfus & Vinner, 1982; Vinner & Dreyfus, 1989). For a fuller discussion of research on learning the concept of function, see chapter 9.

For most students, and indeed for many scientists, the idea of function is completely contained in the "formula". If you ask students for an example of a function, you will often get an algebraic expression such as x^2+3 with no mention of any kind of transformation. Just as with the concept of variable in which the student insists that x "stands for" a *single* number (which may not be known), the concept of function as formula has a very static flavor.

There are a number of ways in which such a function schema is inadequate. For one thing, the objects are restricted to those functions which can be conveniently expressed with a formula. This may suffice for elementary mathematics but it will not do for advanced mathematical thinking. When a function is the same as a formula, the action of evaluation on this object consists of plugging in numbers for letters and composition of two functions is restricted to substitution of a formula for each occurrence of a letter. The notions of domain and range have no place in this schema and graphs, while manageable in themselves (because of their concrete and visual nature), have no connection with functions for the student with a function-as-formula schema. When the graph does not display a clear picture (as is the case with the characteristic function of the irrationals), then the student is unable to think about it.

A more powerful schema for functions will involve interiorization of actions. When a subject perceives a situation that can be dealt with in terms of a function, then we suggest that he or she can view the situation as an action on objects that transforms them into other objects. This action is interiorized. Thus, an important part of what it means to know a function is to construct a certain kind of process that can be used to make sense of a certain kind of phenomenon. Some may refer to this as a mental representation of the function, but we prefer to avoid such terminology because of its tendency towards the misleading suggestion that the internal process is a copy of some "external reality". The important point is that when a function is known as an interiorized process, then this knowledge has a dynamic flavor which affects the nature of the subject's interaction with the function situation.

Evaluation becomes the action of taking a particular value (in the domain of the function) and performing the process on it to obtain a new value (in the range of the function). It may then be possible for the subject to coordinate a function's process and its graph. That is, there is the understanding that the height of the graph of a function f at a point x on the horizontal axis is precisely the value $f(x)$. The subject can then relate to the full power of graphing which is the relationship between the physical shape of the graph and the behavior of the function.

Several important ideas in mathematics can be described as doing some of the things we have discussed with the process of a function. For example, the coordination of two processes and the composition of the functions (see Ayers et al, 1988). A function's process can be reversed, thereby obtaining the inverse function. It is by reflecting on the totality of a function's process that one makes sense of the notion of a function being *onto*. Reflection on the function's process and the reversal of that process seem to be involved in the idea of a function being *one-to-one*.

We have done some preliminary empirical work relative to the points in the preceding paragraph. We find, for example, that students seem to have more difficulty with the concept of one-to-one than with onto. We suggest that the presence of the reversal in one-to-one explains this observation. Similarly on several occasions we have given subjects the following kinds of problems relative to three specific functions, F, G, H. (See Ayers et al, 1988 for details.)

1. Given F, G find H such that $H=F{\circ}G$.

2. Given G, H find F such that $H=F{\circ}G$.

3. Given F, H find G such that $H=F{\circ}G$.

Of course the first is much easier than the other two, and we find invariably that the third is harder than the second. We can suggest an explanation derived from our theory. The first kind of problem seems to require only the coordination of two processes that, presumably, have been interiorized by the subject. The second, however may require that the following be done for each x in the domain of H.

2a. Determine what H does to x obtaining H(x).

2b. Determine what G does to x obtaining G(x).

2c. Construct a process that will always transform G(x) to H(x).

The third kind of problem may be solved by doing the following for each x in the domain of H.

3a. Determine what H does to x obtaining H(x).

3b. Determine value(s) y having the property that the process of F will transform y to H(x).

3c. Construct a process that will transform any x to such a y.

Comparing 2b with 3b (the only point of significant difference), we can see that 2b is a direct application of the process of G whereas 3b requires a reversal of the process of F.

It is perhaps interesting to note that this difference in difficulty (between 2 and 3), which is observed empirically and explained epistemologically, is completely absent from a purely *mathematical* analysis of the two problems. They are, from a mathematical point of view, the calculation of H$\circ$G^{-1} and F$^{-1}\circ$G, respectively, which appear to be problems of identical difficulty. This seems to be another important example in which the psychological and mathematical natures of a problem are not the same (*cf.* p. 113).

Another situation in which relative difficulty can be explained by the requirement of reversing a process occurs in the development of children's ability in arithmetic. According to Riley, Greeno & Heller (1983, p. 157), "Problems represented by sentences where the unknown is either the first (? + a = b) or second (a + ? = c) number are more difficult than problems represented by equations where the result is the unknown (a + b = ?)." Obviously the first two problem types involve a reversal of the process which, in the third type, can be applied directly.

A number of important mathematical activities may require that the function schema be reconstructed at yet a higher level where a function is not only an interiorized process, but as a result of encapsulation, this process can be treated as an object by the subject. One representation that could help with this is the set of ordered pairs (with the "uniqueness to the right" condition) and another is the graph. We refer to chapter 9 for a discussion of some of the difficulties in this connection. In order for a function to be the result of a mathematical activity (such as solving a differential equation or setting up an indefinite integral) it must be an object. Similarly, it seems to us that the elements of a set must be (epistemological) objects and thus, all of functional analysis with its sets and even structured spaces of functions depends on the object nature of a function.

At the same time, and this may be a further reconstruction of the function schema, it seems necessary in many situations that the subject think of a function simultaneously (or at least in rapid succession) as both a process and an object. Consider, for example, the various binary operations on functions such as point-wise addition, point-wise multiplication or composition. In reflecting on the addition of two functions, the subject must see this as a binary operation which takes two objects and transforms them in a new, third object. To actually do this, however, it would seem that the original two objects must be unpacked or "decapsulated" back into processes, these two processes coordinated (by means of "point-wise addition") and the resulting process encapsulated into an object which is the new function that appears as the result of the operation of addition. The same kind of description can be used, as we have indicated above (see page 104), for composition of functions.

As a final example, consider how complex, in these terms, is the following mathematically straightforward statement.

> In the semigroup hom$(G, \circ)$ of endomorphisms of a group G under the operation of composition, the subset of those endomorphisms which are isomorphisms form a group.

From our point of view it seems that to understand this statement (and check that it is true) the subject must think of functions as objects since they form a set, and later a subset, and then understand composition as we have described it to get a firm grasp on hom$(G, \circ)$. Now, in dealing with the group axioms, the cognitive interpretation of function goes back and forth between process and object. The two interpretations must be coordinated in order for the subject to grasp the somewhat subtle idea that the group identity is the identity function and the group inverse of a function is its function theoretic inverse — and this connection is not exactly an accident.

4. IMPLICATIONS FOR EDUCATION

We conclude this chapter with some comments on teaching mathematics in light of the theory we have expounded. Our theory does not have anything to say about the affective aspects of the teaching/learning situation. In particular, we have ignored Piaget's notion of equilibration (1985) which for him was the driving force behind the (re-)construction of schemas. We have also omitted consideration of various issues such as discovery versus guided learning, and large classes versus individual instruction versus small-group problem solving. The main implication for education that our theory has, as far as we have taken it, is that, whatever happens, in or out of the classroom, the main concern should be with the students' construction of schemas for understanding concepts. Instruction should be dedicated to inducing students to make these constructions and helping them along in the process.

We can offer one general conjecture about motivation. Whatever is the mechanism (*le source* according to Piaget & Garcia, 1983) that moves students to make cognitive constructions, to learn, it seems to us to be a very natural human drive, on a par with the drive for food or sex. We admit that this suggestion is inconsistent with the experience of most mathematics teachers, especially at the post secondary level, where students, other than those with obvious talent for mathematics, do not seem to be interested at all. Our conjecture is that this is due to the overall approach in the traditional classroom, where the goal, *as presented and defended by the teacher*, is for the student to develop skills in computational procedures, to display on examinations, and to "get a good grade". For reasons which we will elaborate below, the student cannot learn these procedures through understanding, whereas he or she is presented by the teacher with a conflict-free way out — imitate and memorize. Unsurprisingly, most students accept the offer and take this route. But imitation and memorization do not lead to cognitive constructions and the result is that the students' desire to learn through growth is suppressed. He or she is "turned off mathematics".

Our experience has been that when a student is presented with concepts that he or she is capable of understanding, when the constructions are possible for the student, and if this capability is apparent to the student, then a natural drive to learn, to understand, to construct is released and the level of effort and concentration on mathematical ideas leaves little to be desired. This happens even in the presence of difficulty, when the student is confronted with mathematical problems that her or his existing schemas cannot handle. As long as there is something for the student to think about, as long as he or she perceives that cognitive activity is leading to some sort of growth that could, eventually, lead to a solution of the problem, then there is little difficulty in maintaining the students' interest.

We will present, therefore, some examples of how traditional teaching methods do not relate to conceptual understanding as the theory presented here explains it and close with a few brief words about what directions an alternative approach might take.

4.1 INADEQUACY OF TRADITIONAL TEACHING PRACTICES

If we are correct in our hypotheses that learning involves applying reflective abstraction to existing schemas in order to construct new schemas for understanding concepts, then it is a trivial but critical observation that a schema can not be constructed in the absence of prerequisite existing schemas. Traditional teaching often ignores this. Consider, for example, a lecture on induction which begins, "Today, we are going to learn how to make proofs by induction". This statement assumes that the listener has a "proof schema", that is, he or she is conscious of various methods of proof which could be applied in a given situation and is therefore capable of adding a new one. For any students in the class who do not possess such a schema, the statement is not very meaningful. It gets worse when actual problems, theorems to be proved by induction, are introduced. If a student's function schema does not include functions that deal with transforming integers into propositions, then the very statement of a problem can be meaningless. Many students are probably somewhat bemused when, later, the teacher is roaring the admonition, "You don't prove the statement for every n, you prove the implication from n to $n+1$!" If proof is meaningful at all, it means that you prove *something*. For students who have not encapsulated the process of implication and for whom proposition valued functions of the positive integers are not objects, there may be no "somethings" in that admonition. If such prerequisites are not dealt with, then it is no wonder that the student gives up on trying to understand (he or she does not have the right tools) and, because success on examinations is both essential and possible, looks for something to imitate.

Another kind of difficulty arises with the predicate calculus. For many teachers, understanding the meaning of a statement such as,

> For every function f in A there is another function g in A such that f(g(x))=x for all x

is essentially a language problem, not very different from understanding statements such as,

> Every student in the class has a counselor who will be available to give advice every Monday at 9 am.

But there is much more than language present—in both statements. For the first, according to our theory, the student must have constructed (in her or his mind) a set of functions, interiorized a process of iterating through this set picking an object, iterating again to pick another object, and converting the two objects back to their function processes so that it is possible to iterate once again, this time through the domain of the functions, testing the equality. Only after these constructions are made can the problem be treated linguistically. From our point of view, it is the constructions that provide the essential difficulties, the language aspect being fairly trivial. Similar comments can be made about the second statement which most students have little difficulty in understanding. This is because each construction required to understand the second statement is made naturally, in the course of normal student life and every day experience.

This point about languages, if generalized, suggests to us that the traditional lecture itself, depending largely on linguistic transmission, is not very useful in helping students acquire concepts in mathematics. Mental objects and processes, although they may well exist in the mind of the teacher, cannot be transmitted verbally, or even with pictures, to listeners. It is necessary that the listener engage in active construction.

Another difficulty, related to the problems of imitation, memorization, and verbal transmission arises with examples. It is an article of faith with most mathematics instructors that "lots of examples" must be an integral part of any instructional treatment. It is certainly the case that involvement with examples, whether it be doing exercises or thinking about illustrations and demonstrations, serves to reinforce the concepts that are present in the mind of the subject. We suggest, however, that working with examples may not help very much with the *construction* of concepts. Indeed, we agree with Tall (1986) and it is a major aspect of our theory that understanding mathematical ideas come from sources other than looking at many examples and "abstracting their common features", which is what happens if there is only empirical abstraction. Something more is needed and we suggest that it is precisely the construction aspects of reflective abstraction that we have discussed. It is not clear that more than a very few examples are necessary to construct a concept; in some cases (such as the integers in the initial construction of the concept of a ring) a single example might suffice to induce appropriate reflective abstraction. As we have said, we cannot in this chapter give full consideration to the question of how to induce conceptual learning, but one might well reflect on the contrast between the repetitive examples that seem to be required by conventional wisdom and the single, representative example which so often seems to be in the mind of the mathematician who understands a particular concept. Tall (1986) has referred to this as the *generic example* and it is a promising notion well worth further investigation.

We would go farther in our critical view of repetitive examples and suggest that practice can even be harmful. Yes, the effect of practice will be to reinforce structures that are present. But we would raise the question, what structures are these? Are they part of a student's concept image which conflicts with the concept definition (see Tall, 1977)? Consider what happens when a teacher is explaining, with reference to conceptual understanding, how to solve a certain kind of problem. As we have indicated, the student may not be able to understand the concept behind the method. A general investigation of what drives cognitive development may reveal that whenever a subject is subjected to phenomena, *some sort of construction takes place*. To say that the student does not understand could mean that the student has not and does not construct an appropriate schema for the concept being explained. But if it is the case that something *is* constructed, then it would have to be an inappropriate schema. This result is not inconsistent with what teachers seem to observe in their students after making explanations. What, then, will be the effect of following the explanation with "lots of examples". The inescapable conclusion is that the incorrect interpretations will be reinforced, and teachers will pay a heavy price later on in efforts to correct students' misunderstanding. This may well be a source of epistemological obstacles (Cornu, 1983).

This argument is not sophistry. It is offered as an explanation of a phenomenon in education that seems to be generally recognized, but not very well understood. It seems that Van Lehn (1980) was referring to it quite specifically when he wrote, in a study of the procedural "bugs" observed in students doing subtraction, "When a student has just invented a bug, practice may solidify the bug in memory, thus making remediation more difficult" (p. 47). It is possible that this effect also explains the near impossibility of disavowing undergraduates of various misconceptions observed by Tall (1986), Cornu (1983) and others concerning the concept of limits as well as the persistence, in the face of a variety of instructional treatments, of reversal errors in algebra (Clement *et al,* 1981).

It may be argued that these difficulties can be avoided by giving both examples and non-examples with the examples graded so as to display various features gradually. This could be reasonable, but there are dangers. The decomposition should be based on more than the curriculum developer's understanding of the mathematics. Also, there is no certainty that the student will see the examples in the same way that the instructor did. Finally, this really avoids the issue which is that in order to construct a mathematical idea it is necessary to be mentally active. The really important issues in mathematics education have to do with the nature of this activity and what can be done to foster it.

We do not conclude from this discussion that practice with examples should be eliminated. In addition to reinforcing concepts, they may be important for students to become facile with calculations, to develop a "feeling" that something is wrong, or that it all "hangs together properly". Indeed, it is pure speculation but it may be that practice with a process will tend to induce the subject to encapsulate it. It could be that this is the essential point in the relationship between procedural knowledge and conceptual knowledge (Hiebert, 1986). We do not, therefore reject examples and practice. We only caution the instructor to pay attention to what concepts the students have and what exactly is being reinforced when they are set to do "all the even numbered exercises". It is also important to be aware of the *types* of mistakes that a student makes, how he or she tries to justify an answer (whether it is "correct" or not) or just explain how it was obtained.

4.2 WHAT CAN BE DONE

At this point we must conclude, not, unfortunately, with a prescription for putting things right, but with a brief indication of a research and development program that we are engaged in with the hope of constructing a viable alternative to traditional practice for helping students develop advanced mathematical thinking. There are important connections between what is written here and the ideas found in (Thompson, 1985a; Dreyfus & Thompson, 1985).

Our instructional approach to fostering conceptual thinking in mathematics has four steps.

- Observe students in the process of learning a particular topic or set of topics to see their developing conceptual structures, that is, their concept images.

- Analyze the data and, using these observations, along with the theory we have elaborated in this paper and the designer's understanding of the mathematics involved, develop a genetic decomposition for each topic of concern that represents one possible way in which a subject might construct the concept.

- Design instruction that attempts to move the student along the cognitive steps in the genetic decomposition; develop activities and create situations that will induce students to make the specific reflective abstractions that are called for.

- Repeat the process, revising the genetic decomposition and the instructional treatment, and continue as long as possible or until stabilization occurs (if it does).

To this general description we can add the fact that, in designing instruction, we have found activities with computers to be a major source of student experiences that are very helpful in fostering reflective abstractions. For example, it seems that if a student implements a process on a computer, using software that does not introduce programming distractions (such as complex syntax, constructs that do not relate to mathematical ideas, etc.), then the student will, as a result of the work with computers, tend to interiorize the process. If that same process, once implemented, can be treated on the computer as an object on which operations can be performed, then the student is likely to encapsulate the process. It turns out to be possible to create such opportunities for computer experiences relative to reflective abstractions necessary to construct a wide variety of concepts in mathematics, but that is a topic for another chapter.

We have used this approach to design instruction, with extensive involvement of computers, to help students learn mathematical induction, predicate calculus and many other topics in discrete mathematics. Present efforts are directed towards applying the method to functions and to calculus.

ACKNOWLEDGEMENTS

We would like to thank P. Davidson, T. Dreyfus, H. Sinclaire, L. Steffe, and D. Tall who read and commented on a draft of this chapter.

III : RESEARCH INTO THE TEACHING
AND LEARNING OF
ADVANCED MATHEMATICAL THINKING

The final part of the book is devoted to a wide review of the literature of research in the teaching and learning of advanced mathematics. Much of it comes from the last decade and the task is only just begun. In Chapter Eight, Aline Robert and Rolph Schwarzenberger pause to consider the role of Advanced Mathematical Thinking by looking at the transition from school to university to see if there is a noticeable change. Although there is an increase in complexity and in the need for formal definitions and proof, they find that the intellectual viewpoints already developed by students are often carried over to advanced mathematics with serious consequences in lack of success. In particular the training in elementary mathematics to expect an algorithm to carry out the solution of a problem leads students to seek similar success in contexts where this is no longer appropriate.

In Chapter Nine, Theodore Eisenberg considers the function concept, which is given as a formula or a graph in elementary school, and shows that this proves to be resistant to traditional methods of teaching via formal definitions. Bernard Cornu reveals a crucial example of the discontinuity between elementary and advanced mathematics when he investigates the concept of limit in Chapter Ten. In elementary mathematics there are algorithms for arithmetic, for calculations in trigonometry, for solving equations, but in advanced mathematics a limit usually needs to be calculated by indirect methods which are quite different from the student's previous experience. In particular, the limit, as a process of getting closer, may be encapsulated in terms of an "arbitrarily small quantity" rather than conceived in terms of the definition, leading to serious conflict between concept definition and concept image. The same story continues with Michèle Artigue's review of research into the teaching of analysis in Chapter Eleven and with Dina Tirosh's consideration of the concept of infinity in Chapter Twelve. Here we find explicit conflicts between the concept of infinity in the limiting process and the concept of infinity met in set theory. Often they are kept mentally in different compartments, but when intuitions from one area are brought to mind in an inappropriate context, then conflict is inevitable. At this stage an experiment is reported in detail which is designed to encourage students to reflect on the nature of their beliefs and to reconstruct their knowledge.

Chapter Thirteen focuses on the way in which students build the process of proof. Daniel Alibert and Michael Thomas review both the student's success and difficulties with proofs presented to them through traditional exposition and also look to the possibilities of students engaging in the process of conjecture and debate appropriate for the creation of new advanced mathematical ideas.

Finally, in Chapter Fourteen, we close this review by looking to present research and future use of the computer in advanced mathematics. Ed Dubinsky and David Tall return to the question of the increasing use of computers and the way in which the new technology may be changing the nature of the subject.

CHAPTER 8

RESEARCH IN TEACHING AND LEARNING MATHEMATICS AT AN ADVANCED LEVEL[1]

ALINE ROBERT & ROLPH SCHWARZENBERGER

The seven preceding chapters of this book have examined various aspects of "advanced mathematical thinking"; the six chapters which follow report research into the teaching and learning of specific topics. At this point of transition it is therefore apposite to look back to see in what sense the previous chapters have specified aspects of advanced mathematical thinking which are distinct from mathematical thinking at a more elementary level. In particular, in the first part of this chapter we will address the question:

- To what extent are there aspects of advanced mathematical thinking which are specific to the learning of advanced mathematics at college and university?

We will also take the opportunity to look forward to the specific research to come. We will find that the remaining chapters of the book address themselves mainly to specific topics in advanced mathematics, to study the concept images which students develop and the consequent difficulties which they face in their encounters with the subject. We will therefore spend the second part of this chapter casting the net wider to look at other research in advanced mathematical thinking, in particular to ask:

- What research has been done in the development of advanced mathematical thinking which goes beyond the specifics of the acquisition of individual concepts and the associated learning difficulties?

We shall find that the work to come in the book overlaps with the ideas discussed in earlier chapters. For in studying the learning difficulties of individual students we are gaining some insights into those who are becoming part of the community of advanced mathematical thinkers. Certainly we will find differences between those who study mathematics as a means to an end and those who study mathematics as an end in itself. But since mathematics may be defined as "what mathematicians do", to observe and reflect upon the activities of advanced mathematical thinkers is in principle the only possible way to define advanced mathematical thinking. And to study the difficulties of students will focus on central epistemological problems in the growth of mathematical thought. Conversely the aspects of advanced mathematical thinking discussed in the preceding chapters need to be used in the analysis of particular learning difficulties, if only for the formulation of hypotheses which may be tested empirically.

In this chapter, with an eye on earlier chapters and those to come, we therefore consider the teaching and learning of mathematics at the post-secondary level, that is, in most countries, students aged 18 and above taking specialist mathematics courses at colleges or universities. We concentrate in turn on the two questions formulated above.

---

[1] Thanks are extended to Ed Dubinsky for his initial translation of the draft of this chapter into English.

1. DO THERE EXIST FEATURES SPECIFIC TO THE LEARNING
OF ADVANCED MATHEMATICS?

In an attempt to answer this question we will take a broader view than just the psychological factors which have been considered in the first part of the book to attempt to identify differences between elementary mathematics in compulsory education and specialist mathematics at colleges and university.

1.1 SOCIAL FACTORS

A sociological viewpoint focuses first upon the characteristics of the group of students taking specialist mathematics courses. At first sight it appears that there is a major discontinuity from secondary school education: the students are no longer attending compulsory mathematics courses but are continuing voluntarily after a selection process. One might think that, at least above a certain level, the students might already regard themselves as professional mathematicians or alternatively that some learning difficulties might have been removed by enhanced self-motivation and greater willingness to work.

However, a closer look suggests greater continuity than discontinuity. The students are young adults but are not yet financially independent; they usually have to study a mixture of subjects among which mathematics may not be a first priority and do not usually display an attitude to work much different from that at secondary school. In some cases mathematics is a compulsory pre-condition for studying another subject which is the student's main interest. Teaching methods rarely treat the student as an expert but continue to lay stress on the greater and more accurate knowledge of the professional mathematician. One cannot compare "experts" and students. From a sociological viewpoint, a class of mathematics students at college or university does not look much different from a class in secondary school.

We conclude that it is necessary to look elsewhere for features specific to advanced mathematical thinking.

1.2 MATHEMATICAL CONTENT

From a mathematical viewpoint we see an immediate change in the nature of mathematics being taught. In particular there are more new concepts to teach in less time. Furthermore, from a certain stage onwards there is a greater concentration on a small number of fairly similar mathematical topics with each benefiting from the student's experience in the others, in a manner qualitatively different from anything experienced in earlier years. On the other hand the student is faced by a wider range of possible problems arising in a variety of different contexts which cannot all be discussed in full detail.

This has important consequences. The change in the ratio of quantity of knowledge to available acquisition time means that it is no longer possible for the student to learn all new concepts in class time alone; significant individual activity outside the mathematics class is now an absolute necessity. The concepts themselves are also radically different from the student's previous experience; they often involve not merely a generalization but also an abstraction and a formalization, as outlined by Tall in chapter 1 and Dreyfus in chapter 2.

(See also Robinet, 1984.) The student is required to absorb formalized concepts very quickly, which historically evolved more slowly from a mass of special solutions to special problems by many mathematicians. At the same time the student is expected to adopt new, and often strange, standards of rigourous proof (discussed further in chapter 13).

This formalization involves the abstraction of specific properties which apply now not only to the objects from which they were abstracted but also to any objects which obey the properties. This involves the construction of a new mental object which is different from, and therefore may conflict with, the old objects. It causes the long period of confusion which first year university students meet and is a significant barrier to formal advanced mathematical thinking. It gives rise to a fundamental discontinuity in the difficult transition from elementary to advanced mathematics.

Examples are:

- the concept of *convergence of sequences and series* (completely formalized for the first time by Cauchy in 1827, although the practicalities were already well known); this represents a major generalization and unification.

- The concept of *vector space* which appeared as a formalization in the nineteenth century although specific properties had already been used and understood by physicists in special cases.

- The concept of *group* which was formalized as a useful unification after approximately fifty years of working with special cases of groups of permutations and groups of transformations arising from analysis and geometry.

In each case note that the problems from which these concepts arose in an essential manner are not accessible to students who are beginning to study (and expected to understand) the concepts today: for example, from problems like Césaro's lemma to the definition of convergence via ε–δ methods, or from function spaces to the concept of an infinite dimensional vector space, or from solutions of differential equations to the concept of a group of substitutions.

This leads to an apparent feature that distinguishes advanced mathematical teaching from teaching in elementary mathematics: at this level the teacher is usually obliged to present the notions in a lecture course before getting the students to work with them: there seems to be no question of allowing, or making, the students (re)discover certain aspects of the notions before they are formalized. For example, there does not seem to be a way to make the concept of convergence (in a rigorous sense) accessible to students in which the ε-δ definition is likely to be constructed spontaneously. A lecture course at this level seems to have a specific purpose, quite different from the way in which mathematics developed historically.

Nevertheless we must acknowledge that, far from being a feature specific to advanced mathematical thinking, this may be a result of the unimaginative teaching methods currently adopted in colleges and universities. In the past, similar remarks could have been made of most mathematical instruction at primary or secondary school level. But today, with the advent of more use of concrete materials and computer simulations, and greater emphasis on investigative work in many countries, it is less common to find new concepts imposed prior to the students' ability to solve problems and to construct the concepts for themselves. One may hope therefore that improvements in teaching methods at advanced level may ultimately remove this apparent feature specific to advanced mathematical teaching and lead to a fuller appreciation of the full cycle of advanced mathematical thinking processes mentioned in earlier chapters. In view of the cognitive conflicts involved, such improvements must pay particular attention to the treatment of definitions and to agreement on the acceptability of given levels of proof.

1.3 ASSESSMENT OF STUDENTS' WORK

The teacher of advanced mathematics must also double as an assessor. Therefore much of what is taught must also feature in the assessment process, indeed may be required *by* the assessment process. Such evaluation usually requires the student to be able to correctly state definitions and reproduce correct proofs as well as apply the theory in related problems. Research into concept definitions (chapter 5) underlines that the correct statement of definitions in examinations is liable to degenerate into learning by rote and that many students are often unable to relate directly to the form of the definition. Corresponding research into students' ability to follow or produce proofs (to be reported in chapter 13) confirms that students find proof difficult, with proofs by induction and proofs by contradiction presenting particular difficulty. The only generality on which there is any unanimity seems to be that the required proofs often appear to the student more as a difficult exercise in the use of stylistic set-pieces rather than as an exercise in convincing somebody else of an uncertain result.

It is therefore necessary to question the fundamental premises behind the teaching of advanced mathematics. Is the purpose of learning such a vast quantity of abstract concepts part of a wider scientific, critical and even creative form of advanced mathematical thinking, or is it merely to be able to reproduce learned materials and mechanical skills? Does the assessment process actually reflect the learning that is desired?

It is difficult at this level of complexity, so eloquently formulated in earlier chapters, to decompose the student's activities into elementary tasks. The various aspects of their work are so diverse, yet so interrelated and interdependent, that they will necessarily depend upon components derived from different points in the course which may be widely separated in time. It follows therefore that any short-term assessment of learning is likely to be unreliable and may even be impossible. We conclude that the cumulative character of mathematical knowledge implies that assessment must be long-term. Only in the long run might it be possible to establish accurately any links of cause and effect between instruction and acquisition of knowledge and understanding.

In this respect there seems little distinction between the problems of assessing advanced mathematics and elementary mathematics in primary or secondary school. The experience of short-term assessment in the U.S.A., and the inadequacy of current plans to test pupils at ages 7+, 11+, 14+ and 16+ in England and Wales, suggest that the necessity of long-term assessment actually applies to mathematical thinking at all levels.

1.4 PSYCHOLOGICAL AND COGNITIVE CHARACTERISTICS OF STUDENTS

A psychological viewpoint is more likely to focus on older student's enhanced capacity to reflect on their own activity, as advocated in chapter 2 by Dreyfus and chapter 7 by Dubinsky. According to Skemp (1979) this is a feature of all intelligent behaviour which we surely may assume is common to all of those who advance to college or university, particularly to study mathematics. We could therefore hypothesize that advanced mathematical thinking includes the ability to distinguish between mathematical knowledge and meta-mathematical knowledge (e.g. of the correctness, relevance, or elegance of a piece of mathematics); in addition, that students at this level should carry a substantial quantity of mathematical knowledge, experience of mathematical strategies, and working methods, and be able to communicate them at least at some minimal level. However, an investigation of pupils in the upper section (age 18) of French lycées (Bautier & Robert, 1987) showed that the existence of such abilities was by no means uniform and depended crucially upon the current mathematics teacher of the pupil concerned. No doubt studies of sixth form pupils (age 17–18) in England would yield similar results. We conclude from this that these factors may be relatively easy to change but, to the extent that they help to determine the work habits of students at college and university, they must be explicitly taught.

During the years of secondary schooling students seem to develop preferred methods of approach to solving problems which are relatively stable yet may prove too narrow for a wider range of applications. For example, in geometry some students never use vectors and always employ analytic methods, in analysis some students make systematic use of the graphical approach to solve problems, some rely on formal symbolism and others on numerical methods. Some students systematically attempt to use algorithms to solve problems even when they are inappropriate for the problem in question. A study of students beginning real analysis in the first year of university (Robert & Boschet, 1984) revealed a range of systematic behaviour which was not always correct and showed that good students were often characterized by their versatility in being able to change their mode of approach to suit the particular task. Even in countries where entry to university is restricted to a low percentage of the population, it should be noted that the level of attainment may vary significantly and this difference in ability adds to the divergence in student performance. In England and Wales this is indicated clearly by the wide variation in A-level grades (taken at 18+) attained by students taking mathematics courses. In France the same pre-test on mathematical analysis for students entering university (at age 18+) resulted in a 20% difference between the mean score of students registered at the university and those registered in a class preparing for the entrance competition of the *grandes écoles* for engineering (Robert, 1984).

Later chapters look at specific research into topics in advanced mathematics. This will reveal a wide range of difficulty which confirms the loss of meaning encountered in the work of students at university level. It is as if the great complexity encountered by students causes them to lose all means of control over the material.

To summarize, although the students in question may have full capacity for meta-mathematical reflection, they are often hampered by having too narrow a perspective using particular methods of approach or through having conceptual difficulties with the subject.

1.5 HYPOTHESES ON STUDENT ACQUISITION OF KNOWLEDGE
IN ADVANCED MATHEMATICS

Although advanced mathematical thinking in research mathematicians may be characterized by the full cycle of activity from initial intuitions and conjectures through abstractions to definition and proof, it must be clear from the foregoing discussion that little of these creative activities are found in most advanced mathematical students. It is a reasonable hypothesis to suppose that the various cognitive mechanisms which govern individual learning are not qualitatively different for students from those which apply to younger children. Following Piaget, it is therefore important to stress the central role of individual action at this level, especially through the active solution of problems and the idea that a driving force for the construction of knowledge comes from the process of disequilibrium and re-equilibrium as outlined in chapter 1. Also fundamental is Piaget's notion of reflective abstraction (chapter 7) in the meta-theory of mathematical learning and Vygotsky's hypotheses on the role of communication, with the teacher or with other students, in the formation of personal concepts. From a social psychological viewpoint we also underline the importance of socio-cognitive conflicts, of different conceptions about mathematics, and methods of working in mathematics, as envisaged and expressed by pupils and teachers.

The psychological and cognitive characteristics described in §1.4 above lead to a particular interest in the existence of preferred methods for different students. Our hypothesis is that there is a better prospect of successful mathematical learning for the student with knowledge (however imperfect) in many contexts than for a student having greater knowledge in a single context. This preference for versatile learning is based upon repeated verification of the hypothesis with first year university students studying the beginnings of real analysis (Robert & Boschet, 1984). It was found that the weakest performances in the course of the year were by students having initial knowledge in very few contexts (usually numerical) whereas the more successful students also had initial knowledge in the graphical or even the symbolic context. Such results are easily verified but must be viewed with caution – there is of course a danger that they say no more than that those who succeed on a university course are those who knew the material already. The crucial difference appears to be not the mere existence of prior knowledge but the difference between two very different ways of thinking: the reductive effect of systematically functioning through a preferred context as against the liberating effect of bringing to bear changes of context in problems.

A final hypothesis concerns the possibility of explicitly involving students in their own learning. This might be by helping students to participate consciously in their own learning, that is by helping them to learn how to learn or, on the other hand, by the adoption of a specific and explicit didactic contract by the whole class. It takes account of the reflective capacity of older pupils and is particularly relevant where not all the desired concepts can be approached via meaningful problems.

One can cite here similar ideas in the work of Schoenfeld (1985) who is concerned with problem-solving, and not with the mere acquisition of knowledge. For him, performance in "problem-solving" requires not only knowledge ("resources") but also the use of heuristics much more precise and detailed than that of Polya (1945). This use is governed by a conscious control on what one is in the process of doing and requires compatible conceptions of mathematics. To achieve this, Schoenfeld engages in explicit teaching of heuristics and explains the rules of control (that is, in our terminology, gives instructions of a meta-mathematical kind).

1.6 CONCLUSION

The search for single features which are specific to the learning of advanced mathematics proves to be inconclusive. Many proposed features are seen, on closer examination, to display strong continuity with the learning of mathematics at younger ages. Nevertheless, it seems that, when all these features are taken together, there is a quantitative change: more concepts, less time, the need for greater powers of reflection, greater abstraction, fewer meaningful problems, more emphasis on proof, greater need for versatile learning, greater need for personal control over learning. The confusion caused by new definitions coincides with the need for more abstract deductive thought. Taken together these quantitative changes engender a qualitative change which characterizes the transition to advanced mathematical thinking.

2. RESEARCH ON LEARNING MATHEMATICS AT THE ADVANCED LEVEL

We now turn our viewpoint from looking back at the nature of advanced mathematical thinking and the characteristics of the cognitive nature of the processes to explicit empirical research that has been performed in recent years. For the purpose of this chapter it is useful to distinguish three broad types of research:

- First one can distinguish *research centred upon student's acquisition of specific concepts*. In general the object here is essentially to diagnose those difficulties of students which relate specifically to the mathematical structure of the concepts in question. This involves working "inside" the student, for example, by attempting to define those cognitive processes which are assumed to be involved in the student's acquisition of the concept. Alternatively, the student's failure to acquire the concept may be studied by analysing possible conceptual obstacles in the way.

- Secondly, there is *research which is centred upon the organization of math- ematical content*. Such research may focus upon the sequence of problems and courses offered to the students, on the advantages or disadvantages of particular programmes (e.g. in France for engineers), on methods of making use of the student's own resources to enable them to discover concepts for themselves prior to explanation by the teacher, and so on. Here one tries to adapt to advanced mathematical concepts the work by Piaget on action and by Vygotsky on communication, but again with an analysis appropriate to advanced mathematical learning (e.g. Brousseau, 1988; Douady, 1984).

- Thirdly, there is *research which concentrates upon the external conditions under which teaching and learning take place*: can one discover and recreate environments that are productive of advanced mathematical thought? For example, such an environment might be created by a meta-mathematical style of teaching, or by appropriate explicit contracts between teacher and pupils, or by some combination of these.

2.1 RESEARCH INTO STUDENTS' ACQUISITION OF SPECIFIC CONCEPTS

This is the area that has attracted most research into the development of advanced mathematical thinking. At the beginning of the 1980s, simultaneous research by Cornu in France on the learning of limits (inspired by the work of Brousseau and the earlier ideas of Bachelard on epistemological obstacles), by Vinner in Israel on concept images in the learning of geometry and Tall on cognitive conflict in the learning of limits and continuity, all focussed attention on the mental images conjured up by students which conflicted with accepted mathematical definitions. This research will feature in chapter 10 in the writing of Bernard Cornu. His idea of "spontaneous conceptions" (Cornu, 1981) and Tall & Vinner's joint work on "concept image" were the beginning of a whole sequence of pieces of research on cognitive conflict in advanced mathematical concepts. At the same time Robert (1982a, 1982b) was investigated student conceptions of sequences, distinguishing between responses which were dynamic (with a sense of motion towards the limit) and static (being "close" to the limit, or using a formal ε–δ definition). This contrast between dynamic and static, which also featured in the work of Schwarzenberger & Tall (1978), Tall & Vinner (1981), Cornu (1982) was an early characterization of the process-concept duality which appeared quite separately in the work on conceptual entities (chapter 6) and reflective abstraction (chapter 7). The process-concept duality is at the heart of the difficulties with the function concept in all its complexity, which will feature in the next chapter. The conflicts in the limit concept extend into the topics of more advanced analysis to be discussed in chapter 11 and conflict, in a different way, with the concept of cardinal infinity, to be discussed in chapter 12. Only in chapter 13 does the flavour of the research change from conceptual difficulties with individual concepts to the procedural difficulties with mathematical proof.

We shall leave the discussion of these conceptual difficulties to their proper context in later chapters, here we will dwell in more detail on the second and third broad types of research mentioned above.

2.2 RESEARCH INTO THE ORGANIZATION OF MATHEMATICAL CONTENT AT AN ADVANCED LEVEL

We begin with an example from France of research which falls within the second broad type. Various researchers, in particular Douady (1986), have examined the efficacy, for the development of student's problem-solving abilities, of organizing mathematical content according to the following prescription:

- give a new concept which it is hoped that students will learn, design problems which are undertaken before the formal study of the concept and which contain the possibility of using the concept in their solution.

If possible, the concept should arise in at least two contexts and in one of these the student should have sufficient knowledge or, if the need should arise, insight to make good progress on the problem.

In research which follows this prescription, based on "old" knowledge, the instruction follows in clearly defined stages:

- *explanation* of the role of the concept in problems,

- *institutionalization* through the taught course,

- *familiarization* by means of further problems for reinforcement,

and

- *transfer* to contexts where the concept is not explicitly apparent.

To summarize this cyclic programme, due to Douady, one may say that the knowledge of the pupils begins with old tools available for new objects (*explanation*), which are successively brought into play (*institutionalization*) and become fully available (*familiarization*) to once again be new tools for yet newer objects (*transfer*). To be effective this programme must be realized for a sufficiently large number of concepts and especially for those concepts which cause persistent errors. Thus plans for instruction according to this prescription must be based upon research of the first broad type into the mathematical structure of content and persistent difficulties of students.

A good example of such work is that of Artigue (1987) on teaching the qualitative theory of differential equations which is to be discussed in greater detail in chapter 11. She analyses the relationship between the qualitative approach and the algebraic approach and, using the above prescription, arrives at a programme of teaching in four phases:

Phase 1. Introduction to the qualitative approach and discussion of elementary tools (isoclines, translations, invariance under symmetry and direction of variation of solutions) with the help of simple examples of curve tracing (*explanation*).

Phase 2. Exploitation of the new concepts through examples which require matching of pictures of solution curves to the corresponding differential equations (*institutionalization*).

Phase 3. Comparisons between the methods available for solving problems using the qualitative or the algebraic approach by particular reference to the differential equation

$$\frac{dy}{dx} = (x-2)(y^2-1) \qquad \text{(familiarization)}.$$

Phase 4. Concepts and fundamental theorems of the qualitative theory of differential equations, with proofs based upon pictures of solution curves, dealing with barriers, trapping regions, funnels, attractors and so on (*transfer*).

This teaching programme was given to three groups each containing about 30 students. They were then tested by an examination which included both the algebraic solution of a linear differential equation and the qualitative study of the non-linear equation:

$$\frac{dy}{dx} = \left(\frac{1}{1+x^2}\right)^2 - y^2.$$

We note however that it is practically impossible to obtain valid comparisons of the effectiveness of such a teaching programme and a more traditional course; the assessment items appropriate in the two cases will inevitably differ. Nor is the above prescription adaptable to every mathematical concept in higher education. First, there are some concepts which do not admit a "good" initial problem which can precede the formal study of the concept. Secondly, it may be difficult to find suitable problems from a new domain where the concept is not immediately apparent.

In like manner, Tall (1990) seeks starting points in learning sequences which are meaningful to students at their current state of development yet contain the potential to grow into fully fledged mathematical concepts. His archetypal example is that of "local straightness" which students can visualize in a simple intuitive way, giving meaning to the derivative as the gradient of a locally straight curve, yet containing the seeds of the definition of a differentiable manifold. He calls such a concept a "cognitive root" for a theory, as it is intended to take root in the current cognitive understanding of the student yet grow into a fully fledged formal mathematical notion.

Such cognitive roots also prove hard to find. Indeed, a good cognitive root may often prove to be a process that needs encapsulating as a concept, such examples might include the process of counting as the cognitive root of whole number arithmetic, an input-output process as a cognitive root for function, or the idea of an input-output machine, whose output accuracy can be controlled by controlling the input accuracy, as a cognitive root for a continuous function. All of these examples later require an encapsulation of the process as a meaningful object and then an abstraction of the fundamental properties to give a formal definition requiring the construction of the corresponding abstract object. Such routes from informal sources still need to pass through difficult encapsulation and re-construction phases before being brought to formal fruition.

2.3 RESEARCH ON THE EXTERNAL ENVIRONMENT FOR ADVANCED MATHEMATICAL THINKING

Most research which falls in this third broad type has been concerned with teaching which encourages meta-mathematical reflection. There are, *a priori*, many ways of achieving this.

One method is to teach directly in the manner in which we think that one learns by explaining the importance to discover and learn simultaneously, by making clear the effectiveness of general methods, and by clarifying the basis of the new knowledge. This is exemplified by the Grenoble experiment, described by Alibert in chapter 13, in which scientific debates are held between students in a lecture room. It is designed to encourage the creative, synthetic side of mathematics through conjecture, discussion and argument to give the full cycle of mathematical activity culminating in formal proof that was suggested as the true nature of advanced mathematical thinking in chapter 1. A basic difficulty of this method is that not all students do in fact learn in the same way, so that some way must be found of enabling students to make their own choices and decisions. It is important to define clearly the contract between teacher and students in the first few sessions; this may then be justified by the dynamics of learning which it can generate, and used subsequently as a procedural framework.

The objectives and hypotheses behind such a contract are articulated by Legrand *et al* (1988). The objective is to permit the majority of students to understand the meaning of the algorithms that they are using and to achieve positive ownership of the mathematical concepts which arise. For this the researchers recommend a contract based on the following conditions:

- suppression of indications of truth, especially by the teacher,

- restitution of an official, and approved, status for uncertainty,

- creation of a climate that encourages discussion and debate between students,

- direct engagement in mathematical material by students,

- devolution by the teacher of authority and competence to the students collectively,

- use of complex problems to prevent premature or simplistic algorithms which inhibit concept formation.

Further details will be given in chapter 13.

We note that a very early example of such an approach, in this case at graduate level, was the Texas seminar in point-set topology of R. L. Moore in the 1940s. In this activity Moore also encouraged the students to formulate and prove their own theorems without having formal lectures on the topic. For a personal description, and other remarks on advanced mathematical thinking, see Halmos (1985).

A second way to encourage meta-mathematical reflection is to teach generalized method (Robert *et al*, 1987). By this is meant a set of procedures applicable to a collection of similar problems; thus the use of the method implies a certain class of problems to solve and the fixing of the available tools, techniques and methods of attack. It may also imply the development of a certain number of general ideas such as the usefulness of changing the context, the strategy, or even the formulation of the problem. It may include more specifically mathematical points such as the idea of considering the parameters which arise in a problem as variables, or the idea of looking for invariants which characterize the problem that is being considered. It may include the heuristics of Polya or specific techniques of different conceptual fields (geometry, analysis, ...). The work of Schoenfeld already cited illustrates this point of view, clearly showing the difference between explicit instruction in problem-solving techniques and the more common, but much less effective, implicit instruction based on unspoken imitation. Schoenfeld also demonstrates that it is absolutely necessary to clarify the heuristics of Polya by developing more examples of their use in different contexts and he insists on the necessity of the student's control of their own problem-solving. A similar approach is advocated by Mason *et al* (1982) where the focus is purely on the meta-mathematics of formulating, refining, attacking, reviewing problems and their solutions, using general techniques such as specializing to simple cases, or generalizing through systematic specialization seeking patterns. It seeks to give students confidence through negotiating difficulties when they come up against a seemingly impenetrable barrier and the sense to check insights that come suddenly yet may be flawed.

Thirdly, one can suggest instruction based upon the activities of mathematicians themselves, for example through the study of historical mathematical texts. A difficulty here is the barrier of notation and language as well as the extreme difficulty of many concepts when formulated in their original contexts.

We note that each of these three ways of encouraging meta-mathematical reflection may be suitable in one mathematical context but not in another. None is a prescription for automatic success in any area of advanced mathematics. Moreover, there is no guarantee of successful transfer of such meta-mathematical knowledge from one context to another; no more, in any case, than of successful transfer of mathematical knowledge.

To encourage the transfer of meta-mathematical knowledge it is clearly necessary to create opportunities for such knowledge to be used. It is therefore essential to introduce situations which complete a round trip between meta-mathematical instruction and mathematical experience. For example, Robert & Tenaud (1989) have conducted research on the learning of geometry in the final year (age 18) of the French lycée based upon the following scenario: the students receive explicit instruction in the methods of geometry and, simultaneously, work in small groups on geometric exercises, set without hints, for which the solution is facilitated by bringing these methods to bear. Our hypothesis is that this scenario allows the students to improve their ability to get started on a problem, that it permits the teacher to then usefully continue the instruction in methods, and that all this can serve to accelerate the students' learning of geometry. A similar philosophy, though less carefully structured, lies behind the current research into explicit instruction in investigative methods for secondary school children aged 11-16 in England and Wales.

An additional feature of the research of Robert & Tenaud is that it provides case studies for the use of groupwork to encourage meta-mathematical reflection. They place three to four students in each group, and find that the recent instruction in methods leads to group discussion about what method to choose to get started on the problem. Not only do the students become more aware of the effectiveness of the methodological approach but they become more receptive to further instruction in methods (again, always returning to further groupwork on problems), and they appear to become better able to solve more difficult problems. Thus it is the combination of meta-mathematical instruction, return to appropriate problems, and the use of groupwork which together seem to accelerate the students' learning.

Despite all these positive indications, many questions on the encouragement of meta-mathematical reflection remain unanswered. What is the optimal mix of meta-mathematical instruction and ordinary mathematical instruction? What method of meta-mathematical instruction should be used, and how does this depend upon the particular area of mathematics? Where more traditional teaching has led to particular gaps in understanding, or to misconceptions, how can the use of meta-mathematical instruction avoid the same failure? Again, does the answer to this question depend upon the previous experience of the students?

On the one hand there is a need for much more experimental research to answer all these questions. On the other hand, such research can do no more than provide illuminating case studies, and it might be argued that there are unlikely to be general answers to such questions. Background of the students and the social context of their study are important variables which will vitally affect any results. The setting up of research which requires non-traditional organization of teaching may in any case provoke objections both from the institution, which in many countries works under rigid constraints, and from the students and teachers who have deeply held beliefs about the teaching contract between them. We have seen that such views, at least on the part of students, are changeable. Teachers also can change their views, but may sometimes use alleged rigidity of institutional structures as an excuse not to do so. Added to these difficulties is the problem of objective evaluation and comparison. At the level of the individual student we have already mentioned the need for measures of long-term as well as short-term progress; at the level of different instructional treatments it is necessary, for any valid comparison, to find tasks that are symmetrical with respect to, or "equidistant" from, the teaching and this may not be possible. We cannot count on a clear and convincing general result which will influence the views of unconvinced teachers. We can, however, hope for case studies which follow students' long-term development provided that they are not dispersed too widely. All these caveats explain, perhaps, the still very tentative and hypothetical character of this type of research.

3. CONCLUSION

How can we bring together the discussion of what features, if any, are specific to the learning of advanced mathematics in the first part of this chapter, with the examples of research on the learning of advanced mathematics in the second part?

- One still finds over and over again a certain number of difficulties, mentioned in the first part of this chapter, that are related to the complexity of the contents of advanced mathematics: abstraction and formalization being particular stumbling blocks. All research in mathematical education shows that there seems to be no easy way of avoiding the difficulty of abstraction discussed in earlier chapters by Tall, Dreyfus and Dubinsky. On the contrary it seems essential to develop new ways of approaching it.

- Putting aside the research directed at particular complexity of mathematical content, examples of which are described in subsequent chapters, the remaining research (of the second and third broad types described in the second part of this chapter) does have a common feature: the attempt to change the scientific environment of the students to give them a new and more authentic relationship to knowledge that is more akin to that of experts (i.e. researchers and practitioners) than to that of school pupils. It is here perhaps that we might find a genuine application of advanced mathematical thinking: having available in full the resources of the scientific spirit to control, create, and systematically introduce methods of learning and even, perhaps, of effective research.

FUNCTIONS AND ASSOCIATED LEARNING DIFFICULTIES

THEODORE EISENBERG

1. HISTORICAL BACKGROUND

The function concept has become one of the fundamental ideas of modern mathematics, permeating virtually all the areas of the subject. Yet, despite being a powerful foundation for the final edifice of mathematics organized in a formal Bourbaki style, it proves to be one of the most difficult concepts to master in the learning sequence of school mathematics. In part this is due to the layers of complexity and the numerous sub-notions associated with the concept, for even at the most elementary level functions can be approached in a variety of contexts, and depending upon the approach taken, various difficulties surface from the outset. It has been suggested that the problem of mastering this concept, or any mathematical concept, is simply a task-sequencing problem: provide the student with good exposition and appropriately structured exercises to reveal various aspects of the notion, and students will understand, internalize and master the notion. But this is has proved to be theoretical fantasy. For at this point in time, after millions of dollars have been spent on research in an effort to understand how we acquire mathematical concepts, we still do not know how people learn. It is true that learning takes place, but the precise mechanism by which it occurs is unknown. Indeed, we do not even know how to structure activities to teach children how to order the natural numbers (Sinclaire, 1987). Children eventually learn to do this, and we can roughly identify the stages through which this learning passes. But how it occurs, and identifying and sequencing the essential experiences children must have to acquire this skill, remains a mystery. It is the same with acquiring a deep understanding of functions. Initially this was approached as a problem of task-sequencing in order to give children the foundations for the logic of mathematics, but in this regard, with hindsight, we can see that such an effort was a total failure.

In the 1960s new mathematics movement a succession of influential committees proposed that the function concept be used as a unifying factor in school mathematics.

By grade 6 the word function ... (and the ideas behind it) should be established firmly... .
(Cambridge Conference, 1963, p. 98)

Concepts like ... function ... can be introduced in rudimentary form to very young children, and repeatedly applied until a sophisticated comprehension is built up. We believe that these concepts belong in the curriculum not because they are modern, but because they are useful in organizing the material we want to present. (Cambridge Conference, 1967, p. 10)

A formalist approach to functions soon found its way into the classroom based on the encouragement of influential professors, including Adler (1966), Beberman (1956), Begle

(1968) and Fehr (1966). The warnings of others, including Kline (1958), MacLane (1965), Wilder (1967), and Buck (1970), went initially unheeded until it began to be seen that this supposed logical approach to the curriculum did not work.

At the definition level the function concept can be introduced in a variety of contexts, through arrow diagrams, tables, algebraic description, as a black input-output box, as ordered pairs, etcetera. Of all of these approaches the pedagogically weakest and non-intuitive one seems to be the approach using ordered pairs. Here, a function is defined as a certain sort of set; one which is made up of ordered pairs in which no two ordered pairs have the same first element and different second elements. This seemingly innocent definition proved to conjure up all kinds of logistic and epistemological problems, which incredibly, were often addressed explicitly in some school curricula. For example, logically it might be considered necessary to explain what is meant by an ordered pair and the phrases first element and second element? Norbert Wiener found a way around this problem by defining the ordered pair (a, b) to be $\{\{\{a\}, \emptyset\}, \{\{b\}\}\}$, and variations on this notion of a function were taught at literally all levels of the curriculum, from high school (Kline, Oesterle & Willson 1959) to graduate school (Cohen & Ehrlich, 1963). Commenting on this approach to functions Buck stated:

Experience seems to show that the 'a function is a class of ordered pairs' approach is one which imposes severe limitations upon the student and provides a poor preparation for any further work with functions, either in school or later. (1970, p. 255)

Buck was being kind. Others made sharper criticism of this formalistic approach to school mathematics (Hammersley, 1968), and of the perceived philosophical error of building the new mathematics movement upon a Bourbaki-type foundation prominent in higher mathematics at that time, Thom (1971) and Kline (1970).

Many teachers soon became aware of the limitations of the formalistic approach and used more than one of the aforementioned settings for introducing the concept. This was done with the hope that the notion would be well understood in one context and that this deep understanding would be transferred to the other context settings. It did not happen. Students with a clear idea of function notions in one setting had no idea how they applied in other settings. For example extending the concept of a zero of a function in one variable to functions of two or more variables was beyond most high school students. If transfer of learning occurs between contexts research has shown that the situations have to be very similar, and often the transfer has to be specifically pointed out (Carter, 1970). This has been known to researchers for a long time, and was also familiar to many teachers. Indeed, the inability of students to make "obvious" connections led Sweller (1990) to question whether or not transfer of learning exists in the real world as something more than a theoretical construct. Although some recent studies show that transfer of learning can in general be obtained (Brown & Kane, 1988; Case & Sandleson, 1988; Lehman, Lempert & Nisbett, 1988), it seems that in the domain of school mathematics, situations in which transfers do occur are only epsilons apart from one another; the medium level jumps and quantum leaps which might be hoped for seem to escape all but the most able students.

This phenomenon necessarily led to a multiple embodiment approach – introducing the function concept in a variety of settings with the hope of effecting transfer of learning. Some textbooks developed all new function notions in one base setting (such as a graph, ordered pairs, or table) and then applied them to other settings. Other texts used a smorgasbord approach, letting learners sift-out and build their own concept images and *modus operandi* for internalizing the concepts. (See Harel, 1988.) But in spite of all of these efforts, the function concept was and remains difficult for students to learn (Vinner & Dreyfus, 1989).

2. DEFICIENCIES IN LEARNING THEORIES

A major obstacle in discussing the learning problems associated with functions in particular, and mathematics in general, is that there is no generally accepted theoretical framework as a basis for discussion. Functions are found everywhere in mathematics; the binary operations of ordinary addition and multiplication can be thought of as mappings from $\mathbb{R} \times \mathbb{R}$ into $\mathbb{R}$, the mechanics of solving the standard inequalities problems encountered in algebra and trigonometry can be thought of in a function format, as can many of the standard problems in differential and integral calculus. All techniques, which form a major component of school and university mathematics courses, can be discussed from a functional approach.

Many attempts have been made to build theoretical models for such discussions. But two general types of problems emerge with these models. One is that they are episodic in nature, trying to explain why certain errors occur. But the errors are context dependent – and the explanations necessarily differ from context to context. The second general problem is that when more global schemas are developed, they seem to be too general to be of any use to describe and prescribe remedies to overcome the learning difficulties of specific situations. In short, models of learning are either too specific or too general. Let us look at three models currently in vogue which try to describe how mathematical concepts are learned which have been applied to the learning of functions.

Gagné (1970) developed a theory of learning mathematics which is based within a behavioristic framework. He claims that if learning occurs as a result of instruction, then one can do something after instruction that he could not do before instruction. Learning implies a change in behavior; changes are observable and therefore measurable. Hence, for Gagné, evidence of having learned topic X is being able to perform a specific set of tasks which are related to X. He decomposes topic X into sub-topics and each sub-topic into its pre-requisites. For each sub-topic there is a set of tasks which can be used as evidence that one has mastered that particular sub-topic. A tree of pre-requisites unfolds, and continues to unfold until it can be assumed that all tasks for a particular level are within the knowledge base of the learner. That is to say, Gagné builds a hierarchy of tasks through which one must progress to master a particular topic. The problem here is that such hierarchies rarely coincide with real learning. It may be possible to have all the prescribed sub-topics yet not be able to learn the next stage; the subdivision may be such that the low-level detail may obscure the whole picture. The problem with this theory is that it is task-oriented and says little about the learning characteristics of the student.

Schoenfeld, Smith, & Arcavi (1989), on the other hand, have developed a modus operandi for analysing how mathematical understanding evolves. Their chapter in Glaser's text, *Advances In Instructional Psychology*"... focuses on the changes in the mathematical understandings of one student as she explored ... the graphs of simple algebraic functions in the Cartesian plane ...". It provides a fine-grained characterization of the structure of the student's subject matter understanding, and a description of the nature of the change in her knowledge structures as a result of her interactions with the learning environment (page 1). The essence of their approach is that acquired knowledge must pass through four levels of understanding.

Level 1 is concerned with a macro-organization of knowledge at a schema level; for example understanding that in the equation of the line $L: y=mx+b$, the m and b represent the slope and the y-intercept.

Level 2 deals with compiled knowledge, macro-entities and entailments; for example if $m>0$ the line rises, if $|m|$ is large, the line is steep, the point $(0,b)$ is the y-intercept of the line, etcetera.

Level 3 relates to the fine grained superstructure supporting the knowledge, such as realizing that the slope $(y_2-y_1)/(x_2-x_1)$ of a line through two given points can be thought of as two directed line segments and that when $x=0$ in $y=mx+b$ one gets the y-intercept of the line.

Level 4 is when the limited applications context out, and individuals construct the conceptual atoms that are seen at level 3.

In acquiring a deep understanding of any topic one must necessarily pass through these levels.

Dubinsky's theory of learning adapts aspects of Piagetian theory to the acquisition of mathematical concepts (1988 and Chapter 7 this text). Each individual constructs his own mathematical knowledge through the process of reflective abstraction. The theory is concerned with the way in which processes are *interiorized* to become routinized, *encapsulated* to be considered as concepts, *coordinated* (by following one procedure by another), *reversed* (to be performed in the reverse direction) and *generalized* by being placed in a broader context. The meanings of these notions are given more fully in Chapter 7. Dubinsky believes that an individual's mathematical knowledge is concerned with their tendency to respond to a perceived problem situation by (re-)constructing (new-) schema in an effort to deal with the situation. Learning is episodic. He analyses such episodes and puts a partial ordering on the subject matter to produce what he terms a genetic decomposition, then investigates how the genetic decomposition meshes with the schemata of the student. (In addition to the examples given in Chapter 7, see his paper with Lewin on the genetic decomposition of induction and compactness, 1986).

There are obvious similarities between these theories. Each has aspects of decomposing the subject matter into a learning sequence, in which Schoenfeld *et al* and Dubinsky try to determine a closeness of fit between the content decomposition and the changing schemata in the student's mind. In particular, Dubinsky's formulation proves to be specifically apposite for the function concept because the function is both a process (input - output) and a mathematical object that needs to be treated as a conceptual entity (as described in Chapter 6). Therefore the technique of encapsulation of the function process into a mathematical object seems to be a perfect match for the development of the function concept. Indeed, by getting the students to concentrate on the process of programming functions in an appropriate language which allows the functions (or rather their symbols) to be used as objects, Dubinsky has shown some success in the encapsulation process.

But herein comes the major problem which Dubinsky (1988) explicates:

> It is not possible to observe directly any of a subject's schemas or their objects and processes. (p.7)

If this is the case, then it makes it very difficult to map out how one learns in any terms but general ones. States of knowledge can be observed, but the actual movement from one state to another cannot. As Sinclaire (1987) stated, learning occurs, but when observing its growth, one must be content with seeing it in progressively higher but static states. Because of its complexity it is impossible to observe a continuous development. And it is only possible to infer the cognitive structure of students' conceptualizations through the concept images they evoke in written and spoken communication. So how can one be absolutely sure that the student has an *abstract* function concept, related to the function definition, as opposed to a *generic* function concept that can handle all the tasks in a given context, such as the programming of specific functions?

A major problem with developing Gagnéan type hierarchies and "fine structures" of genetic decompositions is that they can get very complex. The lists of prerequisites which must be mastered for even the simplest of the tasks get unwieldy; the topic gets over atomized – and the whole seems to be much more than the sum of its parts. In a Gagnéan hierarchy the integration of the component steps may not be made and, in many cases, seem to result in the accumulation of isolated skills. So to this point in time, although there are bench marks against which one's knowledge of functions can be measured, the content matter delineates the particular states of one's knowledge. Although a lexicon seems to be developing à la Schoenfeld *et al* and Dubinsky to discuss learning in more general terms, attention is mainly drawn to episodic learning, and to catalog the troublesome areas students have with the function concept. We rarely know why these problems occur, nor do we really know how to guarantee a cure. With this in mind, we now focus on some of the fundamental and documented problems in understanding basic notions of functions.

3. VARIABLES

The role of a variable is imperfectly understood. Although it is the building block for all abstractions in mathematics, its meaning escapes many students. Wagner, Rachlin & Jensen (1984) worked with small but carefully chosen groups of ninth grade students in Athens, Georgia and Calgary, Canada. Collectively, the students represented the whole spectrum of ability levels. The purpose of their study was to investigate learning difficulties in elementary algebra. One of the tasks given was to solve an equation for a particular variable; the investigators changed the name of the variable and the students were asked to solve the "new" equation for the "new" variable. With the solution to the original problem in front of them, one-third of the students resolved the problem from the start. Wagner (1981) found similar results with even older high school students, average age 16+. The students seemed to react to such exercises in one of two ways. Either they would accept the change of variable name and state that the letter makes no difference as long as the numbers stayed the same, or they would regard the change of variable name as producing a completely new problem. Transfer of learning was not there. Wagner related this lack of understanding to Piaget's theory of conservation of variable and she tried to categorize several of the misconceptions. But why didn't transfer occur, especially with such a basic idea? On a large scale, even if 25% or even 15% react as did the older students in Wagner's study, the situation is alarming. Who is at fault? The teacher? The students? The authors? The material? The concept of variable is surely not that difficult to understand – or is it?

Arcavi & Schoenfeld (1987) developed a unit to sensitize teachers to how elusive the concept of variable can be and why their students often have trouble with it. Part of their unit demonstrates the variable meaning of variable. The notion of concept images and concept definitions have been discussed in chapter five and also in several papers (Tall & Vinner, 1981, Vinner, 1983), but the main idea therein is that students develop mental pictures of concepts and definition by circumscription. Exemplars and non exemplars forge the concept, which emerges into being inclusive and exclusive. But "the set of mathematical objects considered by the student to be examples of the concept is not necessarily the same as the set of mathematical objects determined by the definition" (Vinner & Dreyfus, 1989). Arcavi and Schoenfeld's work alerts teachers to the hidden problems of learning the concept of variable itself. It is not as simple as writing a definition on the board. Wagner's research drives this point home; but teachers, who themselves have internalized the variable concept, seem to pay little attention to it. As Freudenthal (1983; p. 469) states:

I have observed, not only with other people but also with myself ..., that sources of insight can be clogged by automatisms. One finally masters an activity so perfectly that the question of how and why [students don't understand them] is not asked anymore, cannot be asked anymore and is not even understood anymore as a meaningful and relevant question.

4. FUNCTIONS, GRAPHS AND VISUALIZATION

Although most students can graph simple functions, they often treat the graph of a function as something external to the function itself and not really part of its essence (Vinner & Dreyfus, 1989). Moreover, they may incorrectly relate to data in graphs of functions they themselves drew. The following problem was on a recent matriculation exam in Israel for students in the least demanding mathematics track.

A circle of radius 8.5 cm is circumscribed about a triangle with angles of 100°,34°, and 46°. Find the radius of the inscribed circle.

It is almost impossible to start this problem without first drawing a sketch. But thousands of the students who took this exam drew the sketch incorrectly, and never referred to it again. Students drew the diagram without taking into account the givens of the problem, which force the triangle to sit in a half-circle. But the odd part in all of this is that it really doesn't matter; drawing the diagram correctly seems to complicate the solution process!

This problem was also given to a group of ten high school mathematics teachers. Eight of them drew the problem incorrectly, and only half of them solved the problem within the 15 minute time span allotted. In another problem, more than 300 students were asked to find the equations of the tangent lines to the circle $x^2 + y^2 = 10$ which pass through the point (5,5). Eighty percent of them, including their instructors, did not draw a sketch for the problem.

Similar observations were obtained when students were asked to find values for a and b such that the line $2x+3y=a$ is tangent to $f(x)=bx^2$ at $x=3$, and when students were asked to solve the inequality $x-3 \geq \sqrt{(2x+9)}$, both were approached analytically – without utilizing the visual interpretation of the givens (Selden, Mason & Selden, 1989).

It seems natural to view many aspects of general mathematics, and functions in particular, in a graphical way. But students simply do not have this concept image of a function. They seem tied to processing information and solving exercises analytically, not visually. This is not a new problem. Historically, the "visual concept image" for a function versus an "analytic characterization" was debated for hundreds of years, with the visual image eventually losing out (Kleiner, 1988).

In the wake of this development, the geometric conception of function is gradually abandoned. A new tug of war soon ensues (and is, in one form or another, still with us today) between this novel "logical" ("abstract", "synthetic", "postulational") conception of function and the old "algebraic" ("concrete", "analytic", "constructive") conception.

The tug of war between these characterizations actually became a three way battle, with the "logical" description entering into the picture too. Today, this conception problem seems to have been settled in favor of the analytic description, at least mathematically speaking. But something much deeper seems to have happened. The nature of mathematics itself seems to have been determined in the process. Consider Hilbert's comment (quoted in Hadamard, 1945):

I have given a simplified proof of part (a) of Jordan's theorem. Of course, my proof is completely arithmetizable, (otherwise it would be considered non existent but, investigating it, I never ceased thinking of the diagram (only thinking of a very twisted curve), and so do I still when remembering it. I cannot even say that I explicitly verified (or can verify) every link of the argument as to its being arithmetizable (in other words, the arithmetized argument does not generally appear in my full consciousness). However, that each link can be arithmetized is unquestionable as well as for me as for any mathematician ... I can give it instantly in its arithmetized form, which proves that arithmetized form is present in my fringe-consciousness.

Hence, although he had an intuitive and visual understanding of the theorem, he did not consider it mathematics unless it was arithmetizable. This point of view has dominated the 20th century mathematics (Davis & Hersh (1986), it is perpetuated in the classroom and it is the view most students have of mathematics in general and functions in particular. Mathematics is analytic in nature – that is the nature of the beast.

Clements (1984) has shown that although the mathematically talented can think visually, they have a strong tendency not to do so. Krutetskii (1976) went even further. He claimed that the ability to visualize is not a pre-requisite for having mathematical talent. This certainly provides evidence for the Hilbertian view that mathematics should be formalistic, even though as in Hilbert's proof of Jordan's theorem, the formalism emerges from an intuitive-visual base. This visual base is often down-played in the classroom, With the emphasis being placed on the analytic formalization of it. For many students this is wrong; they can "see" the mathematics, they just can not formalize it. They need to be taught how to analytically describe their visualizations (Vinner, 1983).

Calculus is a case in point. Tall (1978) and Mundy (1985) have shown that students simply ignore the inconvenient. Ninety percent of first year calculus students simply dropped the absolute value sign in integrating $\int_{-2}^{2} |x| \, dx$. Dreyfus & Eisenberg (1987) have observed that students seldom relate to the graphs they themselves draw. Students sketched a correct graph to evaluate $\int_{-4}^{4} |x^2 + 5|x| + 6| \, dx$ but they simply ignored the sketch when evaluating the integral. Schoenfeld (1985) has made similar observations. The list could go on, but the point is that students do not see elementary functions of a single variable as being inherently tied to a graph. What is worse, neither do their instructors. Dreyfus & Eisenberg (1986) gave the following problem to professors of mathematics,

$$\int_{-3}^{3} \sin(x)[\cos(x) + 3x^2 - x\sin(x)] \, dx,$$

and asked them how they would go about solving the problem. All started by saying that they would try integration by parts or substitution, but not one saw initially that the function is odd, and because of the limits of integration, had to be zero. These professors of mathematics did not see the function as a graph because they failed to spot the oddness of the function which trivializes the solution. Other problems, in which both visual and analytic solution methods were equally likely, were presented to the professors. Their overall tendency was to approach the problems and process the information therein analytically. Thinking visually was foreign to them, so how can their students be expected to develop their visualization skills?

The work of Janvier (1978), Karplus (1979) and Ponte (1984) also point to the fact that students simply do not understand some of the graphs that they themselves drew. Indeed, these studies show that many students cannot interpret graphs. That is, they do not understand the relationship the graph describes between the independent and dependent variables. As an example to illustrate this dichotomy between the graph of a function and the function itself, consider the case of a group of 40 post calculus students asked to find the inverse of a graph which was given in both algebraic and graphical form. Ninety percent of them were able to do this for the algebraic case with 55% being able to justify why their procedure worked. Thirty percent knew the "reflecting through the line $y=x$" technique for the geometric case, but not one could justify the procedure. (And not one knew the technique of flipping the paper and rotating it 90°). The students obtained correct results without understanding why they worked. This concept of the inverse function was divorced from a visual interpretation. Hadar, Zaslavsky & Inbar (1987) have categorized errors like these that students make with functions. Thomas (1969) has tried to identify the ages and stages of development students must be at in order to learn specific function topics. But such categorizations and identifications are only first steps and no one to date, seems to have progressed further. Why these errors and misunderstandings continue to occur with students studying from the newest of curricula, remains a topic for further research.

It is true that a certain type of elementary mathematics cannot be done without a fair amount of visualization skills, and that this precedes most work with functions. Again we look at calculus. Here, for example, students must be able to visualize the common volume of two right cylinders of equal radii which intersect at right angles. They must also be able to visualize an object if it is known that it has a circular base and that every plane passing through it, which is perpendicular to a fixed diameter, generates an equilateral triangle. Exercises such as these form the heart of calculus. But the analytic description of the volume, i.e. the function to be integrated, is the key. Tall (1986a), Blackett (1987), Rival (1987), and Thomas (1988) have shown that students can develop a deep understanding of higher level function concepts by developing them visually. Moreover, they have shown that when the emphasis is placed on the visual development, there is higher retention of them than if they were developed in an analytic framework. But is mathematics the visualization of such situations or is it the symbolization of them? It is with the symbolization of situations that students seem to have trouble not, when applicable, with their visual interpretation. (Leinhardt, Zaslavsky & Stein, 1990).

As evidenced at a debate at a meeting of the Working Group on Advanced Mathematical Thinking, there are two schools of thought among researchers as to whether or not school and university mathematics should emphasize the visual aspects surrounding elementary functions (see, for instance, Dubinsky, 1989; Dreyfus 1991). The majority seemed to be in favor of a pro-visualization stance, but there were eloquent pleas for the symbolic view. Those in favor of visualization considered that functions should be thought of graphically wherever possible and when applicable, operations on them should be thought of in this way too. For example, the simple transformations $f(x)\pm k, \pm kf(x), f(x\pm k), f(\pm kx), |f(x)|, f(|x|), f^2(x), 1/f(x)$, etcetera, should be viewed graphically and that this method of viewing functions can be instilled in students through carefully chosen sequences of exercises. The evidence presented in this chapter suggests that progress will be made only through a major shift in how we approach school and university level mathematics. The Hilbert-Bourbakian view of mathematics has produced generations of semi-literates, in part because the pictures which motivate the proofs and which are behind the big ideas are seldom emphasized in the classroom.

It is likely that the situation can be improved by emphasizing the role played by the visual representation of a function. Ben-Chaim (1985) has shown that visualization skills can be learned and should be emphasized. Cipra's supplemental text for calculus (1983) which builds intuition and visualization skills is a good example of such an approach. It is my belief that the present tendency not to teach in a visual way needs to be reversed, unless of course mathematicians are satisfied with the semi-literate and mathematical phobics produced by the present methods (Paulos, 1988).

5. ABSTRACTION, NOTATION, AND ANXIETY

Functions and their associated sub-concepts have degrees of abstraction. Open any higher level mathematical journal or textbook at random and a page of symbols and formulae will generally appear. Students often scan the page to see if the symbols and formulae are familiar; But far too often they are foreign to them and it soon becomes apparent that a tremendous amount of work will have to be done to understand what is written. Davis &

Hersh (1986, p. 269) label this: the loss of meaning through the intellectual process of mathematical abstraction and, as we have seen in chapter 8, it is a widespread problem amongst students. But abstraction is what mathematics is all about, at least according to Hilbert, the Bourbakists and most university professors. And this abstraction, along with the pace with which it comes in the university classroom, is often the downfall of many students. Obviously there are levels of abstraction. But too often this is not realized in the classroom for the topics have been internalized by the instructor; they have mastered the topic and they expect their students to do it too, in record time. Freudenthal's comments on automaticity (*op. cit.*) are certainly apposite. But this feeling that instructors of mathematics can absorb a page of mathematics as though it were an article in a newspaper is often conveyed to the student. And in many cases, students are discouraged before extending an effort. In the literature this is called math phobia or anxiety and surely enough, it is a cause of student failure. That is, learning fails for affective (emotional) reasons, not cognitive ones. But there is irony in all of this, for math anxiety touches every individual – even instructors of mathematics.

Hard as it is for many students to believe, even mathematicians have entire domains of mathematics with which they do not feel comfortable. For some this may be working in 3-space, for others it may be Galois, probability, or ergodic theory; but every individual has "grey areas" which they often try to avoid. It is only within recent years that mathematicians have recognized this obstacle to learning. The "humanistic mathematics movement" led by Alvin White is devoted to sensitizing instructors to this learning difficulty (White, 1988).

Although there are many facets to mathematical anxiety, notational complexities are often obstacles in preventing understanding of function concepts. Again we meet the problem that it is not the mathematics, but the representation of the mathematics. Indeed, Pimm (1984) posits that a positive correlation will be found between learning in texts which minimize the notation and personalize the presentation. His arguments are persuasive, but heretofore untested. Notational difficulties sneak in everywhere in elementary mathematics, but we will chronicle only a few of the pitfalls associated with initial notions on functions.

1. The $f(x)$ notation itself is confusing because $f(x)$ is usually read to mean x under the function f goes to... In the early 70's a movement led by Howard Fehr (1974) used the standard algebraists notation of $x \xrightarrow{f}$ but the movement never caught on. Nevertheless, this notation captures the dynamic aspects of a function which should initially be emphasized. Perhaps then students will have less trouble with understanding composite functions of the form $x \xrightarrow{g} g(x)$, $g(x) \xrightarrow{f} f(g(x))$). Moreover, this notation should further help students understand the meaning of the argument. E.g., seventy percent of beginning calculus students at Ben-Gurion University could not solve the following problem:

If 2 and 4 are the values of x for which $f(x)=0$, what are the values of x for which $f(4x)=0$?

This same problem was rephrased:

Only the values of 2 and 4 go to zero under the function f. What values multiplied by four will go to zero under the function f ?

Seventy percent of the students could answer the problem in this form, and they seemed to have a basic understanding of what they were doing. This is a common phenomenon even in simple word problems: that the phrasing of the question greatly affects the students ability to answer it.

Students often do not realize that functions transform every point in the domain to a new position. Until this is understood, problems such as finding the zeros of f(kx) given the zeros of f(x) cannot be fully understood.

2. Realizing that $\int \dfrac{du}{u\ln(u)}$ is of the form $\int du/u$ is already evidence of looking at functions in a more general way, although even this may only involve manipulation of symbolic formulae. But *defining* a function in terms of an integral such as in $\ln(x) = \int_1^x dt/t$ is beyond most students in elementary courses – a common difficulty in the first stages of advanced mathematics where the idea of definition, rather than description, is so new. Likewise, visualizing functions in parametric form, also proves to be intolerably difficult, especially when the representation moves from two dimensions to three.

Functions can be thought of as representing forms, and students need a wide set of experiences in looking at functions in this way. Moreover, this inability to work with concepts like functions, as though they are objects is a major obstacle for many students, many of whom have demonstrated an ability to work with lower level abstractions in the past.

As Harel and Kaput have already asserted in chapter 6, the isomorphism between a vector space and the dual of its dual in linear algebra provides fertile ground to illustrate this learning difficulty. A linear transformation from a vector space V to the vector space $\mathbb{R}^1$ is called a linear functional. The set of all linear functionals from V to $\mathbb{R}^1$ forms a vector space called the *dual* of V; this vector space is usually denoted as V^*. Here, the vectors in V^* are functions, which are combined in the usual way. But the dual of V^*, is also a vector space, V^{**}, which is isomorphic to V. Students have a very hard time in understanding the role of the vectors in V^{**}. Certainly they are functions. But functions do things to *objects* in a domain. What do these functions do to the functions in V^*? The functions in V^{**} must send the functions in V^* somewhere, but to where? As has been discussed in chapter 6 in terms of conceptual entities and chapter 7 in terms of encapsulation, it is very difficult for students to grasp that the functions are acting as operators as well as objects. This dual role is difficult for students to understand because of the layering of the abstractions. And higher level mathematics is full of such situations.

The problem solving research of Major & Clark (1963) has shown that when experts are presented with a problem that they are seeing for the first time and which is not even close to their domain of expertise, they approach a solution in the same way as novices. That is, there is a playing around period, and generally an incubation period before realizing a solution. For example, consider the following two mathematical problems:

1. For each real number x, let f(x) be the minimum of the numbers $4x+1$, $x+2$, *and* $-2x+4$. What is the maximum value of f(x)?

2. There is a kingdom where if a person drinks poison he will die. The only way to counteract the poison is to drink a stronger one. Then the reaction stops. The king decides that he must have the strongest poison available in his possession. So, he sets up a contest between his court adviser and his wizard. Each must find the strongest poison in the kingdom and give it to him. But to be sure that he will get the strongest poison, he will force each to first drink the others' poison. Then they will drink their own. One of them will die, but the king will certainly then have the strongest poison in his possession.

The court adviser knows that he can never out smart the wizard, and bemoans the fact that in a few days time he will die. But soon, he and his wife think of a plan to outsmart the wizard.

The day of the contest arrives. The wizard drinks the adviser's poison, and vice-versa. Then they drink their own and the wizard dies.

What was the adviser's trick?

These problems were given to novice and experts in mathematics. Both groups were observed trying to set up functional relationships to describe the various situations in problem two and to substitute values into problem one. But the two groups essentially attacked the problems in the same way. Polya's (1954) and Schoenfeld's (1985) map of problem-solving tactics was followed almost to the letter, including the fact that not one of the interviewees wanted to spend more than two minutes on the problem. This, Schoenfeld claims is characteristic of most problems in the curriculum, from grade school through the university. Interestingly, most of the interviewees felt the second problem wasn't mathematics, in spite of their symbolic-logistic solution paths. There is relevance in these problems for our discussion on functions.

In order to solve the above problems a reversal tactic must be used. This is exactly the same sort of skill which must be used to find $g(x)$, given $f(g(x))$ and $f(x)$, or $f(x)$, given $f(g(x))$ and $g(x)$. Moreover such reversals are necessary when determining, for example, the

function under consideration if it is known that the integral $\pi \int_0^1 [(1-x^2)^2 - 1^2]\, dx$

represents the volume generated when a curve is revolved around a certain line and you must determine the volume generated when this curve is revolved around the y-axis. Such problems are very difficult for students, even though they often have all the skills to work out the solution. They need time to internalize the skills involved in working with functions in this way. Freudenthal's admonitions on automaticity must always be in our minds; concepts and topics are not self-evident to students, even after our lucid explanations of them.

6. REPRESENTATIONAL DIFFICULTIES

Birkhoff (1956) developed a metric for aesthetics which showed that a mathematical object's appeal, be it a function, axiomatic system or whatever, is inversely proportional to its complexity. How one measures this is another matter. But it seems obvious that the more symbols, signs, *etcetera*, the more complex – and everyone, not just students, has trouble with complexity.

$$\lim_{r \to \infty} \lim_{m \to \infty} \lim_{n \to \infty} \sum_{\mu=0}^{m} (1 - (\cos(\frac{(\mu!)^r \pi}{x}))^{2n})$$

is a function which gives, for positive integers x the largest prime factor of x (Boas, 1960).

Notation often causes problems with students, even though, like many of the problems discussed above, the basic underlying ideas which they represent are simple. Anxiety, the nature of mathematics and the loss of meaning through the "intellectual" process of over ambitious use of symbolism and abstraction have been discussed above. Good mathematics is not necessarily complex mathematics, and complex mathematics is not necessarily good mathematics. But, given the choice, it seems obvious that one would opt for the less complex and the more intuitive. Unfortunately, this has not been the option chosen in the past.

7. SUMMARY

In this chapter we have surveyed some of the learning difficulties students have with function concepts, and why they occur from historical and psychological points of view. Perceived helplessness and anxiety seem to hinder learning, and literature has been cited which studies these psychological barriers. But the major theme of this chapter is that functions and their associated notions are not conceived visually, and that this non-visual approach hinders one's development of having a sense for functions. Students seem to think of function concepts in only a symbolic representational mode. Indeed, not only are functions thought of this way, but this seems symptomatic of the fact that the majority of topics constituting mathematics are considered non-visual. It is the conclusion of the author that this unwillingness to stress the visual aspects of mathematics in general, and of functions in particular, is a serious impediment to students' learning.

CHAPTER 10

LIMITS[1]

BERNARD CORNU

The mathematical concept of a limit is a particularly difficult notion, typical of the kind of thought required in advanced mathematics. It holds a central position which permeates the whole of mathematical analysis – as a foundation of the theory of approximation, of continuity, and of differential and integral calculus.

One of the greatest difficulties in teaching and learning the limit concept lies not only in its richness and complexity, but also in the extent to which the cognitive aspects cannot be generated purely from the mathematical definition. The distinction between the definition and the concept itself (discussed in detail by Vinner in chapter 5) is didactically very important. Remembering the definition of a limit is one thing, acquiring the fundamental conception is another. One facet is the idea of approximation, usually first encountered through a dynamic notion of limit, and the way in which the concept of limit is put to work to resolve real problems which rely not on the definition but on many diverse properties of the intuitive concept. Starting from such a point of view students often believe that they "understand" the definition of a limit without truly acquiring all the implications of the formal concept. Students are often able to complete many of the exercises they are asked to perform without having to understand the formalism of the definition at all. Meanwhile, the quantifiers "for all", "there exists", which occur in epsilon-delta definitions, have their own meanings in everyday language subtly different from those encountered in the definition of the limit concept. From such beginning arise conceptual obstacles which may cause serious difficulties.

In teaching mathematics, certain aspects of the limit concept are given greater emphasis which are revealed by a review of the curriculum, the textbooks, exercises and examinations. In the first half of the twentieth century, French mathematics texts used the notion of limit in an intuitive manner without a formal definition to introduce the definition of the derivative. Later in the same text a definition would be given which is more in the manner of an "explanation" in a note at the foot of the page. The official French programme of school mathematics first cited the term limit with respect to the derivative as long ago as 1947. In 1966 the notion was properly introduced into the programme. Books generally devoted a chapter to the general limit concept including a formal definition, a statement of its uniqueness, and theorems about arithmetic operations applied to limits. The exercises, however, did not concentrate on the limit *concept*, but on inequalities, the notion of absolute value, the idea of a sufficient condition and, above all, on *operations*: the limit of a sum, of a product, and so on. These exercises are far more related to algebra and the routines of formal differentiation and integration than to analysis. They take on such an overwhelming importance that one textbook cited thirty one different theorems on operations on limits!

---

[1] The author wishes to thank Rebecca Tall for her initial translation of the draft of this chapter into English, and the editor for help and assistance in completing the task.

Given such a bias in emphasis it is therefore little wonder that students pick up implicit beliefs about the way in which they are expected to operate.

Different investigations which have been carried out show only too clearly that the majority of students do not master the idea of a limit, even at a more advanced stage of their studies. This does not prevent them from working out exercises, solving problems and succeeding in their examinations!

In this chapter we will study some didactic aspects of the idea of limits: concepts linked to this notion, various obstacles which stand in the way of students learning the limit concept, and discuss various strategies for teaching the limit concept.

1. SPONTANEOUS CONCEPTIONS AND MENTAL MODELS

For most mathematical concepts, teaching does not begin on virgin territory. In the case of limits, before any teaching on this subject the student already has a certain number of ideas, intuitions, images, knowledge, which come from daily experience, such as the colloquial meaning of the terms being used. We describe these conceptions of an idea, which occur prior to formal teaching, as *spontaneous conceptions* (Cornu 1981, 1983). When a student participates in a mathematics lesson, these ideas do not disappear – contrary to what may be imagined by most teachers. These spontaneous ideas mix with newly acquired knowledge, modified and adapted to form the students personal conceptions. We know that in order to resolve a problem, we do not in general call uniquely on adequate scientific theory, but on natural or spontaneous reasoning, which is founded on these spontaneous ideas. This phenomenon is well-known in the empirical and theoretical development of scientific concepts since Bachelard in the nineteen-thirties, but it is only in the last decade that it has been fully realized that exactly the same forces operate in the apparent logic of mathematics.

In the case of the limit concept, we observe that the words 'tends to' and 'limit' have a significance for the students before any lessons begin (Schwarzenberger & Tall, 1978), and that students continue to rely on these meanings after they have been given a formal definition. Investigations have revealed many different meanings for the expression 'tends towards':

- to approach (eventually staying away from it)
- to approach ... without reaching it
- to approach... just reaching it
- to resemble (without any variation, such as "this blue tends towards violet")

The word limit itself can have may different meanings to different individuals at different times. Most often it is considered as an 'impassible limit', but it can also be:

- an impassible limit which is reachable,
- an impassible limit which is impossible to reach,
- a point which one approaches, without reaching it,

- a point which one approaches and reaches,
- a higher (or lower) limit,
- a maximum or minimum,
- an interval,
- that which comes 'immediately after' what can be attained,
- a constraint, a ban, a rule,
- the end, the finish.   (Cornu, 1983)

From one student to another the meaning given to words varies; for one student it may have many meanings, according to the situations. Spontaneous ideas live on a long time; investigations show that they may remain with students at a much more advanced stage of learning. In the face of a variety of spontaneous notions and the student's growing awareness of the formalisms it easily happens that contradictory ideas may be held simultaneously in the mind of an individual, leading to a global "concept image" which contains potential conflicting factors in the sense of Tall & Vinner (1981), as discussed in chapter 5.

Aline Robert (1982a,b) has studied different models which students may hold of the notion of the limit of a sequence. Despite the fact that students have been given a formal definition of a sequence, when asked to describe the notion of a sequence, like as not they would be liable to evoke conceptions relating to various aspects of their previous experience. Some students suggested primitive, rudimentary models, reminiscent of those which might be evoked spontaneously, such as:

- *stationary*: "The final terms always have the same value",
- *barrier*: "The values cannot pass l".

In addition there were more models which arose more from the formal teaching:

- *Monotonic* and *dynamic-monotonic* :
  "a convergent sequence is an increasing sequence bounded above (or decreasing bounded below)";
  "a convergent sequence is an increasing (or decreasing) sequence which approaches a limit".
- *Dynamic* :
  "u_n tends to l";
  "u_n approaches l";
  "the distance from u_n to l becomes small";
  "the values approach a number more and more closely".
- *Static* :
  "The u_n are in an interval near l";
  "the u_n are grouped round l";
  "The elements of the sequence end up by being found in a neighbourhood of l".
- *Mixed* : a mixture of those above.

Once more she found these models influencing the manner in which students at university solved problems. There is clearly no single notion of limit in the minds of students. It is evident that the students have a variety of concept images.

Moreover, it is also evident that the initial teaching tends to emphasize the *process* of approaching a limit, rather than the *concept* of the limit itself. The concept imagery associated with this process, as exemplified above, contains many factors which conflict with the formal definition ("approaches but cannot reach", "cannot pass", "tends to", etc). Thus it is that students develop images of limits and infinity which relate to misconceptions concerning the process of "getting close" or "growing large" or "going on forever".

In an ethnographic study of the conceptions of students concerning limits and infinity, Sierpińska (1987) analysed the concept images of 31 sixteen year-old pre-calculus mathematics and physics students. She then classified them into groups which she labelled with a single name for each group:

Michael and Christopher are unconscious *infinitists* (at least at the beginning): they say "infinite", but think "very big". ... For both of them the limit should be the last value of the term ... for Michael this last value is either plus infinity (a very big positive number) or minus infinity. ... It is not so for Christopher who is more receptive to the dynamic changes of values of the terms. The last value is not always tending to infinity, it may tend to some small and known number.

George is a *conscious infinitist*: Infinity is about something metaphysical, difficult to grasp with precise definitions. If mathematics is to be an exact science then one should avoid speaking about infinity and speak about finite numbers only. In formulating general laws one can use letters denoting concrete but arbitrary finite numbers. In describing the behaviour of sequences *the most important thing is to characterize the nth term* by writing the general formula. For a given *n* one can then compute the exact value of the term or one can give an approximation of this value.

Paul and Robert are *kinetic infinitists*: the idea of infinity in them is connected with the idea of time. ... Paul is a *potentialist*: To think of some whole, a set or a sequence, one has to run in thought through every element of it. It is impossible to think this way of an infinite number of elements. The construction of an infinite set or sequence can never be completed. Infinity exists potentially only. Robert is a *potential actualist*: it is possible [for him] to make a "jump to infinity" in thought: the infinity can potentially be ultimately actualized. For both, Paul and Robert, the important thing is to see how the terms of the sequence change, if there is a tendency to approach some fixed value. For Paul, even if the terms of a sequence come closer and closer so as to differ less than any given value they will *never reach it*. Robert thinks *theoretically* the terms will *reach it in the infinity*.

In this way she exhibits timeless conflicts about limits and infinity which have been with us since time immemorial and which continue to hold in our students today.

Other limiting processes, such as the concept of continuity, differentiation, integration, and so on, whilst on the surface being very different, in cognitive terms exhibit similar difficulties. For instance, continuity suffers from having a spontaneous conception that is evoked through the use of everyday language in phrases such as "it rained continuously all day" (meaning there was no break in the rainfall) or "the railway line is continuously welded" (meaning that there are no gaps in the rail). This viewpoint is often reinforced by teacher's attempts to give a simple insight into the notion of continuity by speaking of the graph "being in one piece" or "drawn without taking the pencil off the paper", thereby confusing the mathematical notions of continuity and connectedness.

A questionnaire administered to first year university mathematics students (Tall & Vinner 1981) included the a question to investigate the students' concept images of continuity (figure 16).

Mathematically f_1, f_2 and f_3 are continuous, whilst f_4 and f_5 are not. But the students' concept images suggest otherwise (Table IV – "correct" responses in bold print).

Although all the responses to f_1 are "correct", the majority are "right answers for wrong reasons", such as the idea that f_1 is continuous "because it is given by only one formula". The function f_2 often causes dispute even amongst seasoned mathematicians. It is continuous according to the ε–δ definition *on the domain* $\{x \in \mathbb{R} \mid x \neq 0\}$. But the students'

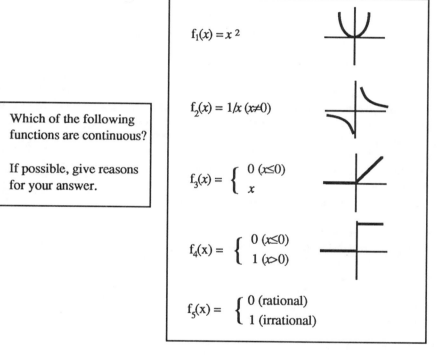

Figure 16 : the concept image of continuity

N=41	f_1	f_2	f_3	f_4	f_5
continuous	**41**	6	**27**	1	8
discontinuous	0	35	12	**38**	**26**
no response	0	0	2	2	7

Table IV : Responses to figure 16

concept images suggest:

It is continuous because:
    "the function is given by a single formula".

It is not continuous because
    "the graph is not in one piece",
    "the function is not defined at the origin",
    "the function gets infinite at the origin".

In the initial stages of learning, we therefore see spontaneous conceptions arising which are often in conflict with the formal definition.

2. COGNITIVE OBSTACLES

The notion of a *cognitive obstacle* is interesting to study to help identify difficulties encountered by students in the learning process, and to determine more appropriate strategies for teaching. It is possible to distinguish several different types of obstacles: *genetic* and *psychological* obstacles which occur as a result of the personal development of the student, *didactical* obstacles which occur because of the nature of the teaching and the teacher, and *epistemological* obstacles which occur because of the nature of the mathematical concepts themselves. In planning to teach a mathematical concept it is of the utmost importance to determine the possible obstacles, particularly the endemic epistemological obstacles.

The term was introduced by Gaston Bachelard (1938):

"We must pose the problem of scientific knowledge in terms of obstacles. It is not just a question of considering external obstacles, like the complexity and the transience of scientific phenomena, nor to lament the feebleness of the human senses and spirit. It is in the act of gaining knowledge itself, to know, intimately, what appears, as an inevitable result of functional necessity, to retard the speed of learning and cause cognitive difficulties. It is here that we may find the causes of stagnation and even of regression, that we may perceive the reasons for the inertia, which we call epistemological obstacles."

He goes on to say:

"We encounter new knowledge which contradicts previous knowledge, and in doing so must destroy ill-formed previous ideas."

He indicated that epistemological obstacles occur both in the historical development of scientific thought and in educational practice. For him, epistemological obstacles have two essential characteristics:

- they are unavoidable and essential constituents of knowledge to be acquired,

- They are found, at least in part, in the historical development of the concept.

Many authors have become interested in epistemological obstacles. Guy Brousseau defines an epistemological obstacle as knowledge which functions well in a certain domain of activity and therefore becomes well-established, but then fails to work satisfactorily in another context where it malfunctions and leads to contradictions. It therefore becomes necessary to destroy the original insufficient, malformed knowledge, to replace it with new concept which operates satisfactorily in the new domain. The rejection and clarifying of such an obstacle is an essential part of the knowledge itself; the transformation cannot be performed without destabilizing the original ideas by placing them in a new context where they are clearly seen to fail. This therefore requires a great effort of cognitive reconstruction.

3. EPISTEMOLOGICAL OBSTACLES IN HISTORICAL DEVELOPMENT

It is useful to study the history of the concept to locate periods of slow development and the difficulties which arose which may indicate the presence of epistemological obstacles. In the case of the history of the limit concept, we see that this notion was introduced to resolve three principal types of difficulty:

- geometric problems (area calculations, consideration of the nature of geometric lengths, "exhaustion"),

- the problem of the sum and rate of convergence of a series,

- the problems of differentiation, (which come from the relationship between two quantities which simultaneously tend to zero).

There are four major epistemological obstacles in the history of the limit concept:

1) *The failure to link geometry with numbers.*

When the Greeks became interested in mathematics about 400-300 BC, we must ask why it happened that the limit concept was not clarified at the time. The problem of calculating the area of a circle, for example, supplied an opportunity to develop the tools very similar to the limit concept. Hippocrates of Chios (430 BC) wanted to prove that the ratio between the area of two circles is equal to the ratio of the squares of their diameters. He inscribed regular polygons within the circles and, by indefinitely increasing the number of sides, he approached the areas of the two circles. At each step the ratio of the areas of the inscribed polygons is equal to the ratio of the squares of the diameters, and it followed that, "in the limit", it would be true also for the areas of the circles.

This passage towards the limit, very sparingly explained, would be defined a year later, in terms of the method of exhaustion, credited to Eudoxus of Cnidos (408-255 BC). The method is based on the principle of Eudoxus (Euclid's *Elements*, book 10, proposition 1) that "given two unequal lengths, if from the first is taken a part larger than its half, then from the remainder a part more than half what remains, and the process is repeated, then there will come a time when what remains will be less than the second length". In other words, by successive halving we can attain a size as small as we wish. From this the principle of exhaustion follows which allows us to state that for any $\varepsilon > 0$ there exists a regular polygon inscribed within a circle whose area differs from that of the circle by less than ε. If the ratio of areas of two circles is A_1/A_2 and that of the squares of the radii is r_1^2/r_2^2, then we have one of three possible cases:

$$A_1/A_2 < r_1^2/r_2^2, \ A_1/A_2 > r_1^2/r_2^2 \text{ or } A_1/A_2 = r_1^2/r_2^2.$$

We eliminate the first two by the principle of exhaustion, and hence deduce the truth of the desired equality.

However, despite the fact that the exhaustion method seems extremely close to the notion of limit, we cannot affirm that the Greeks possessed the modern limit concept. The method of exhaustion is in essence a geometrical method which allows the proof of results without having to deal with the problem of infinity. It is applied to geometrical magnitudes, not to numbers. Each case is dealt with on an individual basis using a specific argument tailored to the geometrical context. There is no transfer from geometrical figures to a purely numerical interpretation, so the unifying concept of limit of numbers is absent. The geometrical interpretation, and its success in resolving pertinent problems, is therefore seen to cause an obstacle which prevents the passage to the notion of a numerical limit.

2) The notion of the infinitely large and infinitely small.

Throughout the history of the notion of limit we meet the supposition of the existence of infinitesimally small quantities. Is it possible to have quantities which are so small as to be almost zero, and yet not having a specific 'assignable' size? What happens at the instant when one of these quantities becomes zero? Such philosophical problems have occupied the attention of numerous mathematicians who, like Newton, spoke of the "soul of departed quantities" at the time that they disappear to enable him to calculate their "ultimate ratio". Euler freely used the notion of the infinitely small as a quantity that can, where appropriate, be considered equal to zero. D'Alembert opposed the use of infinitely small quantities and sought to remove them from the differential calculus. He reasoned that a quantity is either something or nothing. If it is something it cannot be made zero and if it is nothing it is already zero. Thus the supposition that there is an intermediate state between the two he described as a wild dream.

Cauchy also used the language of the infinitely small. In his *Cours d'analyse de l'Ecole Polytechnique* of 1821, he defined a continuous function in these terms:

The function f(x) is continuous within given limits if between these limits an infinitely small increment *i* in the variable *x* produces always an infinitely small increment, f(x+i)–f(x), in the function itself.   (as translated in Boyer, 1939, p. 277)

He explained the idea of an infinitesimal as follows:

One says that a variable quantity becomes infinitely small when its numerical value decreases indefinitely in such a way as to converge to the limit zero.   (quoted from Boyer, 1939, p. 273).

For Cauchy an infinitesimal is simply a variable which tends to zero.

The idea of an 'intermediate state' between that which is nothing and that which is not is frequently found in modern students. They often view the symbol ε as representing a number which is not zero yet is smaller than any positive real number. In the same way individuals may believe that 0.999... is the 'last number before 1' yet is not equal to one. There is a corresponding belief in the existence of an integer bigger than all the others, yet which is not infinite.

3) The metaphysical aspect of the notion of limit.

The notion of limit is difficult to introduce in mathematics because it seems to have more to do with metaphysics or philosophy. Mathematicians are often reticent in speaking of such concepts, from the time of the Greeks through to D'Alembert who wrote "One can quite easily do without the rest of all this metaphysics of the infinite in the differential calculus". Lagrange expressed a similar horror of the metaphysical aspects. Although in his early career he believed he could make the use of infinitesimals rigourous, he later considered that the infinitesimals of Leibniz has no satisfactory metaphysical basis and recast the foundations of the calculus using infinite series in purely algebraic terms. However, this too proved elusive, for

When Lagrange endeavored to free the calculus of its metaphysical difficulties, by resorting to common algebra, he avoided the whirlpool of Charybdis only to suffer wreck against the rocks of Scylla. The algebra of his day, as handed down to him by L. Euler, was founded on a false view of infinity. No rigorous theory of infinite series had been established.   (Cajori, 1980, p. 257)

In this way, whichever way mathematicians seemed to turn in the historical development of the subject, they came against profound theoretical difficulties.

The metaphysical aspect of the notion of limit is one of the principal obstacles for today's students. In an interview one said, "It is not really mathematics", because the initial stages of calculus no longer rely purely on simple arithmetic and algebra. The students may have difficulties handling the concept of infinity, "It isn't rigourous, but it works"; "it doesn't exist", "it is very abstract", "the method is all right, provided you are content with an approximate value". This obstacle makes the comprehension of the limit concept extremely difficult, particularly because a limit cannot be calculated directly using familiar methods of algebra and arithmetic.

4) Is the limit attained or not?

This is a debate which has lasted throughout the history of the concept. For example, Robins (1697-1751) estimated that the limit can never be attained, just as regular polygons inscribed in a circle can never be equal to the circle. He asserted "We give the name ultimate magnitude to the limit which a variable quantity can approach as near as we would like, but to which it cannot be absolutely equal". On the other hand, Jurin (1685-1750), said that the "ultimate ratio between two quantities is the ratio reached at the instant when the quantities cancel out", "it is not a question whether the increment is zero, but that it is disappearing, or on the point of vanishing", "there is a last ratio of increments which vanish", "an increment born is an increment which starts to exist from nothing, or which begins to be generated, but which has yet to attain a magnitude that may be assigned to such a small quantity". D'Alembert insisted that a quantity should never become equal to its limit: "To speak properly, the limit never coincides, or never becomes equal to the quantity of which it is the limit, but is always approaching and can differ by as small a quantity as one desires".

This debate is still alive in our students. In a discussion one asked, "When n tends to zero, isn't n equal to zero?" The following dialogue between students clearly illustrates the epistemological obstacle:

- the more n grows, the more $1/n$ approaches zero.

- as much as one would like?

- no, because one day they will meet.

There are certainly many other obstacles to the notion of limit other than these four. The mistakes which students make are valuable indications for locating obstacles. The construction of pedagogical strategies for teaching students must then take such obstacles into account. It is not a question of avoiding them but, on the contrary, to lead the student to meet them and to overcome them, seeing the obstacles as constituent parts of the revised mathematical concepts which are to be acquired.

4. EPISTEMOLOGICAL OBSTACLES IN MODERN MATHEMATICS

It is interesting for mathematicians to look back at history and note the struggles that gave birth to modern ideas, leading to the logical state of the art today. However the twentieth century quest for certainty based on a secure axiomatic foundation begun by Hilbert foundered on Gödel's incompleteness theorem, and so uncertainty remains. In chapter 4, Hanna has shown that acceptance of proof remains strongly dependent on peer approval. The introduction of Weierstrassian analysis, depending only on logical definitions of number concepts failed to eliminate the infinitesimal concepts that were an essential part of earlier mathematical culture. Although we may formulate definitions of limits and continuity in epsilon-delta terms, we still have occasion to use dynamic language of "variables tending to zero" in a manner analogous to that of Cauchy, with the resultant mental imagery linked to the "arbitrarily small".

Cognitively this phenomenon is to be expected. The idea of an "arbitrarily small" number is but the *object* produced by the encapsulation of the *process* of getting small in terms of Dubinsky's theory of encapsulation (chapter 7). As Tall hypothesizes (chapter 1), the formation of a mental concept of an "arbitrarily small number" is a *generic limit concept* where the encapsulated object is believed to have the properties of the objects in the process. Thus the generic limit of a set of numbers which tend to zero is an arbitrarily small, yet non-zero, number. The concept is a natural consequence of the way in which the mind is hypothesized to work.

Hence, despite the attempts at banning infinitesimals from modern analysis, it continues to live in the minds and communications of professional mathematicians, even if it was eliminated from formal proofs. The return of the logically based infinitesimal in the work of Robinson (1966) re-opened the debate, which continues to be hotly contested. Although Robinson thought that his neat logical solution would solve the three hundred year conflict, this proved not to be so. For Robinson's construction of a hyperreal system containing real numbers and infinitesimals depends on a version of the axiom of choice and is therefore non-constructive. This is becoming more a bone of contention as the arrival of computers begins to focus mathematicians on the pragmatic need to provide finite algorithms for construction of concepts. For instance, the intermediate value theorem is seen to be constructive but the existence of a maximum value of a continuous function is not. The former asserts the existence of a zero of a continuous function between two places where the function has opposite signs and can be programmed on a computer by a simple bisection argument, but the latter depends essentially on a non-constructive proof by contradiction.

In this way we see a recurrence of the problem of Lagrange as he attempted to remove the metaphysical ambiguity from the calculus: just as difficulty seems to be resolved, another seems to appear to take its place. This is typical of the complexity of the ideas in analysis and of the fundamental limit concept.

That the limit concept is essentially difficult may be seen in the way that it is defined in terms of an unencapsulated *process*: "give me an $\varepsilon>0$, and I will find an N such that ..." rather than as a *concept*, in the form "there exists a *function* $N(\varepsilon)$ such that ..." This means that the proof of the first theorem on the algebra of limits (that the sum, product etc of the limit of two sequences is the sum, product etc of the limits) is framed in process terms as the coordination of two processes, rather than as the combination of two concepts. Were the latter to be the case, then the proof would follow a similar format, but it would have the advantage that it could be programmed on a computer in such a way that the proof of continuity is merely the operation of a computer algorithm. Yet this unencapsulated pinnacle of difficulty occurs at the very *beginning* of a course on limits presented to a naïve student. No wonder they find it hard!

5. THE DIDACTICAL TRANSMISSION OF EPISTEMOLOGICAL OBSTACLES

Given the complexities of modern mathematics and the cultural colourations in meaning, it is no surprise that such complex interactions affect students in their learning. In their human interactions they are very sensitive to tone of voice and implicit meanings and such ideas are conveyed to them by their teachers. Although such meanings may be avoided in written texts, they can passed on inadvertently from generation to generation as the teacher tries to "simplify" the complexities to "help" the students. When Orton (1980) investigated the limit concept in terms of a "staircase with treads", he showed a student the picture in figure 17 and asked

(a) If this procedure is repeated indefinitely, what is the final result?

(b) How many times will extra steps have to be placed before this "final result" is reached?

(c) What is the area of the final shape in terms of "a", i.e. what is the area below the "final staircase"?

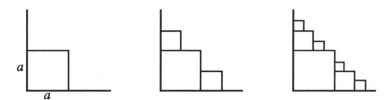

Figure 17 : A limiting staircase

If a student gave a formula in response to (c) he asked:

Can you use this formula to obtain the 'final term' or limit of the sequence ?,

His justification for using this terminology was that:

The expression "final term" was again used in an attempt to help the students understand the meaning of limits.

However, in the light of what has been said in this book about generic limit concepts, it is evident that a phrase such as "the final staircase", far from helping the students with the formalities, is likely to create a generic limit concept in which the student imagines a staircase with an "infinite number of steps". This is precisely the response that it evoked.

In such ways, despite all our attempts to help students through the complexities, our attempts to "simplify" can lead directly to the cognitive obstacles which we have described earlier.

Such obstacles are almost certainly essential parts of the learning process. Davis & Vinner (1986) suggest that there are seemingly *unavoidable* stages in which misconceptions are bound to occur, in line with our assertion that such obstacles require a cognitive re-construction which are bound to involve a period of conflict and confusion. They too highlight the misconceptions that arise from the use of language evoking inappropriate images in spontaneous conceptions. Even though they attempted to teach a course in which the word "limit" was not used in the initial stages, they concluded that "avoiding appeals to such pre-mathematical mental representation fragments may very well be futile". They observed that another problem arises from the sheer complexity of the new ideas which cannot appear "instantaneously in complete and mature form" and so "some parts of the idea will get adequate representations before other parts". They give evidence, substantiating the discussion of Robert and Schwarzenberger in the previous chapter, that specific examples dominate the learning, so that when monotonic sequences feature heavily in the student's earlier experiences, it is not surprising that they dominate the students' concept images.

6. TOWARDS PEDAGOGICAL STRATEGIES

The diversity of conceptions, the richness and complexity of notions, and the cognitive obstacles makes the teaching of the limit concept extremely difficult. Numerous attempts have been made and the problem remains unresolved! On considering these attempts, it is possible to focus on certain fundamental points and to pose essential questions.

In the first place, far too many teachers seem to consider that it is sufficient to present a clear exposition of the limit concept to enable the students to understand. It is far more important that the students are made aware of the complexity of the notion and to reflect on their own ideas and epistemological obstacles. Research so far shows clearly that the students own conceptions are very varied, that they make fundamental mistakes and that they do not necessarily overcome epistemological obstacles. It is necessary for teacher education to take place to help teachers become aware of the problems involved. It is equally important for students to become explicitly aware of the essential difficulties. Experiments have been carried out in which, before starting a session on the notion of limit, the students were given appropriate activities to help them become aware of their own spontaneous ideas, images, intuitions, experiences which they possessed before and which would necessarily be brought into play during the learning process. In particular, they were made aware of the different meanings of the words which they were going to use. This proved to be a valuable technique and enabled them to build on their own knowledge and understanding (Cornu, 1983; Robert, 1982a).

A further problem is that of the context in which the learning takes place. An effective apprenticeship needs to take place in a problem-solving context. The notion of limit has to be used to solve specific problems. It is therefore necessary to present situations in which the student can see that the limit is a useful tool, in which the limit is seen as part of the answer to questions which the student may have asked for him (or her)self. This is often lacking in contemporary teaching. A definition of the notion of limit is given, followed by a sequence of problems and exercises, usually based solely on handling the algebra of the limit concept: the limit of a sum, of a product, of the composition of two functions or of two series ... We have already seen just how difficult the unencapsulated logical form of the limit definition is to handle for experts, let alone beginners.

It is important to consider the order in which the limit concepts are presented. Not only is there the question of designing a logical mathematical order of concepts, but also the cognitive appropriateness of the curriculum sequence and of the problems to be solved. It is now well-established that in the transition to advanced mathematical thinking a purely logical sequence of topics, in which the mathematical concepts are introduced through definitions and logical deductions, is likely to be insufficient.

Some alternative methods of approach will be discussed in later chapters. Tall (1986), for example, decided that the research evidence on students' difficulties with the limit concept made it totally inappropriate to approach it either through the formal definition or even (in the case of the derivative) through a geometric experience of a secant "tending to a tangent". He hypothesized that the limit process should be used *implicitly* in the calculus as part of a magnification process to "see" the gradient of a curve and gain experience of the concept in action before the concept becomes the focus of formal discussion. In this sense he is following a similar path to that of Douady (1986) in her "tool-object" dialectic in which the concept is first used implicitly and informally as a tool to gain appropriate cognitive experiences before it becomes the explicit focus of attention as a mathematical object. (Douady's ideas are also discussed in §2.1 of chapter 8 and Tall's graphic approach to limits in the calculus will be discussed in the next chapter).

Dubinsky (chapter 7) has formulated the notion of a "genetic decomposition" of mathematical concepts, that is to say a collection of reflective abstractions which are approached in a certain order to provoke the learning of the concept envisaged. Thus if we want to introduce the concept of the limit of a series we must first try to describe the concept image and the sequence of mental constructions which are necessary for the student to make. As students with different experiences and different cognitive structures are unlikely to require exactly the same sequence, one may hypothesize that it is essential that the students actively and consciously participate in the reflective abstraction process to reconstruct their knowledge structure and build the limit concept.

As we shall see in chapter 14, the computer may very well play a significant role in providing an environment where the student may gain appropriate experiences to construct the limit concept. However, such approaches are very likely to contain their own peculiar epistemological obstacles (Tall & Winkelmann 1988) and it is necessary to reflect deeply on student experiences in the new environment to see precisely what is learnt and in what form the knowledge is held in the mind. The interaction with the computer may involve programming, for the individual to construct computer processes which, through reflection, may permit the acquisition and mastery of the corresponding mathematical constructs. It may involve pre-prepared software to enable the student to experience carefully selected environments which model the idea of a limit. It is equally possible to imagine a kind of computer "toolbox" for the learning of the limit concept: a computer environment which will permit the students to manipulate objects and to construct knowledge: to recognize and construct sequences, to operate on them, constructing new sequences, transforming and manipulating them, studying their behaviour and the nature of their convergence.

Various other approaches are possible. In a context such as that of studying limits it is vital that the computer software is designed within a teaching strategy based on the careful analysis of the concept due to be acquired. Spontaneous conceptions, concept images, obstacles, reflective abstraction, and genetic decomposition, are all conceptual tools designed to assist in the design of such pedagogical strategies.

CHAPTER 11

ANALYSIS

MICHÈLE ARTIGUE[1]

The conceptual field of analysis is vast. At the elementary level it is structured around the notions of:

- real number,
- function,
- limits of numerical sequences and functions,
- continuity,
- the derivative and integral of functions of one real variable.

At more advanced levels these extend to analysis of several variables, complex analysis, functional analysis, measure theory and so on.

The two preceding chapters have summarized empirical research and cognitive theory relating to the first four of these headings. This chapter concentrates on the smaller body of work focussing on the fundamental notions of differential and integral calculus.

In the initial section we will briefly review major points in the historical evolution of the concepts and the ways that they have been taught. In the second we will use empirical research and theoretical interpretations to draw up a catalogue of the mental conceptions constructed by students engaged in traditional education. Finally, in the third, we will present some instructional treatments taking this research into account and designed to improve student understanding.

---

[1] Thanks are due to Ed Dubinsky for his initial translation of the first draft of this chapter and to the editor for his assistance with the final version.

167

1. HISTORICAL BACKGROUND

1.1 SOME CONCEPTS EMERGED EARLY
BUT WERE ESTABLISHED LATE

It is well known that the fundamental notions of differential and integral calculus appeared on the mathematical scene very early, but their development was very slow.

From the time of antiquity, calculations of length, area and volume, based on the method of exhaustion, opened the way to integral calculus. In the seventeenth century, the problems of tangent, maxima and minima, linked in particular to the study of celestial mechanics and ballistics in their turn opened the way to differential calculus. This set the scene for the independent development of the calculus by Newton and Leibniz, culminating in the reciprocity of the operations of integration and differentiation. The first text-book: *Analysis of the infinitely small for understanding curved lines* was published by the Marquis de l'Hospital (1696).

It was not until the beginning of the nineteenth century that Cauchy developed a firmer theoretical basis for the calculus using the notion of limit, and integration was developed using continuous functions. In the remainder of the nineteenth century the arithmetization of analysis was carried out, through formal definitions of the real line by Dedekind cuts or Cauchy sequences, and formal definitions of limits and continuity using $\varepsilon \angle \delta$ methods in a purely arithmetic form by Weierstrass. This led Boyer (1939) to claim:

the unequivocal symbolism of Weierstrass may be regarded as effectively banishing from the calculus the persistent notion of infinitesimal.

Meanwhile, it was not until 1893 that Stolz introduced the notion of differentiability for functions of several real variables and only in 1911 that the development of functional analysis led to Fréchet introducing the differential in its modern interpretation in terms of linear tangent maps.

The latest twist in the story occurred in the 1960s, when Robinson formulated a rigorous theory of non-standard analysis, reintroducing infinitesimals on a logical basis after a century of rejection.

1.2 SOME CONCEPTS CAUSE BOTH ENTHUSIASM
AND VIRULENT CRITICISM

It is well known that from its birth, infinitesimal calculus has excited passions. On the one hand there is the enthusiasm of those who, like the Marquis de l'Hospital, are astonished by the possibilities opened up by the algebraisation of calculus:

The extent of calculus is immense: it is as easy for mechanical curves as for geometrical ones; it is indifferent to radical signs and even makes use of them; it extends to as many variables as one would wish; comparisons of all kinds of infinitesimals are equally easy. Furthermore, an infinity of surprising discoveries has come out of it.   (preface to de l'Hospital 1696)

On the other there is virulent criticism from those for whom infinitesimals are beings without roles, carriers of paradoxes, the manipulations of which are based on dubious practices. Thus Berkeley fiercely criticizes the arguments of Newton:

This reasoning seems neither correct nor honest. For when one says that increments are no longer anything or that there are no more increments, the preceding supposition to the effect that increments were something or that there were increments is destroyed, yet a consequence of the supposition is retained... This is a false reasoning.

Likewise, D'Alembert wrote (in his article "Differentiel" in the *Encyclopedie Methodique*):

It is not at all a question of how one speaks ordinarily of infinitesimal quantities in differential calculus: it is just a question of limits of finite quantities. Thus the metaphysics of infinity and some infinitesimal quantities being larger or smaller than others, is totally useless in differential calculus. One only uses the term infinitesimal to abbreviate expressions. We would not say therefore, as do many geometers, that a quantity is infinitely small neither before it vanishes nor after it is vanished, but in the very instant at which it vanishes: for wouldn't that mean a very false definition, a hundred times more obscure than that which one wishes to define?

A little later Lagrange (1797) would attempt to liberate analysis simultaneously from both limits and infinitesimals, judging each of the two approaches to be as subject to criticism as the other.

1.3 THE DIFFERENTIAL/DERIVATIVE CONFLICT
AND ITS EDUCATIONAL REPERCUSSIONS

At the heart of these differences of opinion is the differential/derivative conflict, originally a debate between the English school using Newton's fluxions and the continental school using the differential of Leibniz. On the continent during the eighteenth century, the differential of Leibniz was one of the essential motors of the development of differential and integral calculus.

But the structuring of differential calculus around the notion of limit led to the progressive decline of infinitesimals and in their wake differentials were supplanted by partial derivatives (defined in terms of the limit concept). The differential survived in analysis – reduced to the role of a formal expression invariant with respect to a change of variables and therefore a useful tool for calculation and memorization. It survived also in applications, especially in physics, with a role approaching its original status under Leibniz – as an infinitesimal increase – a useful tool for substitution in equations, but suspected of being based on less rigorous practices.

The article "Differential Calculus", in the French version of the German Encyclopedia of Mathematical Sciences, gave a very good account of the situation at the end of the 19th century. The differential was defined only at the end, in reference to work of Cauchy, as the product of the derivative by an arbitrary increase of the variable. The author accompanied his definition with the following comment:

In fact the Leibnizian notation is *theoretically* superfluous. *Practically*, it has a great importance...
In applications, usage is facilitated if one agrees, once and for all that one understands by the symbol
dy not the differential f'(x) dx but a very small increase in the function y=f(x)... A *differential formula*,
that is, a relation between x, y, dx and dy, coming before expressing for example, the law of a physical
phenomenon, no longer has, it is true, *any precise meaning*. It is, however, easy to establish and there
is no inconvenience in using it, for in order for it to acquire a precise meaning, it suffices to divide
both sides by dx and pass to the limit as dx tends toward 0...

(Voss, 1899, translated from the French edition p. 279)

It is only in the course of the twentieth century that the differential reappeared, in the
development of functional analysis, as the key notion of local approximation, but this time
with a subtly different role: that of a tangent linear functional.

These debates and conflicts, and the atmosphere of scientific uncertainty that they
engendered, reverberated throughout education for an even longer time.

Differential and integral calculus was introduced into secondary education at the
beginning of the century in many countries and in 1913 the Commission Internationale pour
l'Enseignement des Mathématiques (CIEM)[2] organized a study of the subject, published
by the journal "Enseignement Mathématique" the following year. The general reporter
wrote:

Scientific literature itself has not made a clear resolution of the diverse definitions of differential.

(Beke, 1914, p. 272)

The atmosphere of uncertainty still persisted in 1930:

This question (differentials) is one of the points in mathematics where those who search for precision
do not always find, even with the best authors, the desired clarity. Prudent recommendations as to
the mode of using differentials of higher order which generally accompany the exposition, doubtless
permit correct *calculation* with differentials: but the goal of education is more than that.

(Delens, 1930, p. 333)

To this is added the problem of developing the most rigorous possible instruction, without
metaphysics and therefore without infinitesimals, and to eliminate errors by beginners
based on formal and automatic manipulation of differentials influenced by their status
during this period. In an article in the same journal, on definitions in mathematics, Poincaré wrote:

As soon as one passes to derivatives of the second order, one swims in absurdity: let z be a function
of a variable y which is itself a function of x; I write:

$$\frac{d^2z}{dx^2} = \frac{d^2z}{dy^2} \times \frac{dy^2}{dx^2} + \frac{dz}{dy} \times \frac{d^2y}{dx^2} \ .$$

In this formula I write d^2z twice, and the symbol has two different meanings...

---

[2] The CIEM later became ICMI (International Commission for Mathematical Instruction).

The difficulty is aggravated if one has several independent variables. I write:

$$dz = \frac{dz}{dx} \times dx + \frac{dz}{dy} \times dy.$$

Here again we have three occurrences of the symbol dz with three different meanings...

How can we avoid these traps? Beginners will not be able to do so.   (Poincaré, 1899, p. 107)

The author cites a delightful story, where a student came to the equation of propagation of sound, and after having simplified dt^2, extracted the square roots of differential elements by simply suppressing the indices "2":

$$\frac{d^2x}{dt^2} = a^2 \Rightarrow \frac{dx}{dt} = \pm\, a$$

Beke concluded the paragraph of his CIEM report dedicated to differentials in these terms:

We are, I believe, unanimous in desiring that the metaphysical haze of infinitesimals should not enter into secondary education. I am of the opinion that the wisest method is to not introduce differentials at all in secondary education. This view is justified by the actual tendency to eliminate differentials in the whole of mathematical science. How much more necessary would it appear to reject the teaching of all notions which give rise to so much misunderstanding.

Thus in secondary education, teaching differential calculus was based on the notion of derivative, defined as a limit of a quotient and associated with the geometric picture of the tangent as a limiting position of secants. In higher elementary education, preference was given to derivatives and partial derivatives, whilst differentials were defined in terms of these and limited to first order.

Even so, the arguments between supporters and opponents of differentials continued to be virulent (see, for example, Laurent, 1899). In the *Mathematical Gazette*, controversy raged for several years following the appearance of an article of E.G. Phillips (1931) regretting that students might come to the university without having heard a mention of differentials. He proposed to introduce them from the beginning of "Elementary Calculus" using the modern definition of differentiability arising from the increase of the function proportional to the increase of the variable. There followed a debate on this question at the annual meeting of the Association in 1934. Subsequently the derivative was introduced as:

$$\frac{dy}{dx} = \lim_{\delta x \to 0} \frac{\delta y}{\delta x}$$

where $\delta y = f(x+\delta x) - f(x)$ for a small value of δx. The symbol $\frac{dy}{dx}$ now has the status of a single indivisible symbol where dy and dx are given no individual meaning:

dy/dx must, at least for some considerable time, be regarded as an inseparable whole, just as δx is. It does not in any simple or straightforward way mean anything like 'dy divided by dx' and a statement such as 'dy/dx $\times$ dx/dt = dy/dt by cancelling dx' is just so much gibberish.

(*SMP Advanced Mathematics Book I*, 1967, p. 221)

1.4 THE NON-STANDARD ANALYSIS REVIVAL
AND ITS WEAK IMPACT ON EDUCATION

The publication in 1966 of Robinson's book *Non-Standard Analysis* constituted, in some sense, a rehabilitation of infinitesimals which had fallen into disrepute since the arithmetization of analysis. By using a logical construction based on ultrafilters, he proposed a rigorous foundation for the approach to differential and integral calculus using the infinitely small and the infinitely large. It was met with suspicion, even hostility, by many mathematicians who saw it only as a useless reintroduction of discredited, even dangerous, archaic tools whose rejection had done nothing to hinder the development of mathematics for more than a century. Nevertheless, despite the obscurity of this first work, non-standard analysis developed rapidly both in mathematical research and in research into logical foundations. In the latter case a major goal was that of simplifying the initial construction of Robinson or to propose axiomatic approaches (for example, Nelson, 1977).

The attempts at simplification were often conducted with the aim of constructing an elementary way of teaching non-standard analysis. This was the case with the work of Keisler (1971, 1976) and Henle & Kleinberg (1979). The first work of Keisler served as a reference text for a teaching experiment in the first year of university in the Chicago area during 1973–74. Sullivan (1976) used two questionnaires to evaluate the effects of the course: one designed for teachers, the other for students. The eleven teachers involved gave a very positive opinion of the experience. The student questionnaire revealed no significant difference in technical performance between standardists and nonstandardists, but showed that those following the non-standard course were better able to interpret the mathematical formalism of calculus and to make sense of it.

The appearance of the second book by Keisler (1976) led to a virulent criticism by Bishop (1977) in the *Bulletin of the American Mathematical Society*, accusing Keisler of seeking the goal of modern mathematicians: to convince students that mathematics is only "an esoteric and meaningless exercise in technique", detached from any reality. These criticisms were in opposition to the declarations of the partisans of non-standard analysis who affirmed with great passion its simplicity and intuitive character. For example Henle and Kleinberg wrote in the preface to their work:

Thus we were led to the ε–δ approach to calculus, an approach that, although totally precise and rigorous, was a disaster for students to learn and teachers to teach... A most natural place for Robinson's insight is a next (and possibly final) point in the evolution of the teaching of calculus. We can now develop calculus using infinitesimals and enjoy all of their simplicity and intuitive power, yet at the same time work in a mathematically precise and rigorous atmosphere.

We shall return to this question at the end of this chapter. However, it is necessary to emphasize the weak impact of non-standard analysis on contemporary education. The small number of reported instances of this approach are often accompanied with passionate advocacy, but this rarely rises above the level of personal conviction.

1.5 CURRENT EDUCATIONAL TRENDS

As the majority of research cited in the remainder of this chapter is from France and England, we restrict ourselves to very brief descriptions of recent educational trends in these two countries.

Calculus/Analysis Teaching in France:

After the reform of 1902, the derivative was introduced in secondary education to students aged 16–17. In 1971 the classical definition, in terms of the limit of a quotient of differences, gave way to a definition in terms of affine approximation, the derivative appearing as a by-product of the approximation – the coefficient of the linear part, as evidenced by the following extracts from the curriculum:

"Linear function tangent at a point to a given function; derivative at this point..."   (1971)

"Expansion limited to order 1; derived number, dynamic interpretations (velocity) and geometric interpretations (tangent)..."   (1982)

"Approximation by an affine function in a neighborhood of 0, functions which associate to a given h: $(1+h)^2$, $(1+h)^3$, $1/(1+h)$, $\sqrt{(1+h)}$. When, for a neighborhood of $h=0$, $f(a+h)$ can be written in the form $f(a+h)=f(a)+Ah+h\varepsilon(h)$ with $\lim \varepsilon(h)=0$ when h tends to 0, one says that the function f has A for its derived value at a..."   (1985)

Correspondingly, the tangent is presented as the straight line of best local approximation to the curve associated with the function. But, since the reform of 1982, the $\varepsilon\angle\delta$ formalization of the limit has been omitted.

Integration is now introduced in the last year of secondary education (age 17–18), traditionally defined in terms of the primitive, whose existence is assumed for a function continuous on an interval. The calculation of primitives is immediately applied to the calculation of areas (area of a domain in the plane defined in a rectangular coordinate system by the relations $a \le x \le b$ and $0 \le y \le f(x)$, f being a continuous positive function); it is specified in the syllabus that the difficulties involved in the notion of area will not be introduced.

The reform of 1972 also introduces a more ambitious program in the domain of integral calculus, with the definition of Riemann sums for a numerical function of a real variable on a bounded interval; the theorems on the integrability of continuous or piecewise monotone functions are admitted.

The reform of 1982 sees the return of the integral as primitive and as the area under a positive function, and introduces examples of approximating the value of an integral by various numerical methods.

Differential equations in the syllabus are only concerned with algebraic solutions in the most simple cases. The latest programs only mention linear differential equations with constant coefficients of the first and second order without a second term.

At university, in mathematics or physics, formal instruction in analysis constitutes the major part of the first two years. In differential calculus, the derivatives, partial derivatives and Jacobian matrices occupy centre stage. The notion of differential is introduced at the beginning of the study of functions of several variables. For the last twenty years this has been in terms of its modern role in the tangent linear map. Integration is concerned with the classical development of the Riemann integral. Differential equations are taught, but essentially only with the goal of obtaining algebraic solutions.

Calculus/Analysis Teaching in England:

In England prior to the Education Reform Act of 1989 there were no national directives on the curriculum and this act only concerns itself with education to the age of 16. The syllabuses at "advanced level" in school (aged 16–18) are determined by external examinations which are offered by a variety of competing examination boards who decide their own content subject to the agreement of advisors from the teaching profession and the universities, subject to an agreed "common core". The pure mathematics content of the various syllabuses is based on the algorithms for differentiation, integration and simple ideas about differential equations. The concepts are explained in a dynamic way ("as x tends to a" or "$x \to a$") and the course is mainly concerned with the *methods* and *applications* of differentiation and integration; only a very few may see the ε–δ definitions at a later stage. The methods of the calculus may be applied in other areas such as mechanics. At university, mathematics students study the logical foundations of analysis using ε–δ definitions (or equivalent topological formulations), whilst other students study calculus methods or analytic theory to a level appropriate for their main subject. Formal analysis is known to trouble all but the most able mathematics students and in some universities there is a trend to reduce the formalities of the subject and concentrate more on methods and applications.

2. STUDENT CONCEPTIONS

One can associate some *a priori* concepts with the notions of derivative, integral, tangent and tangent plane. For example, one can conceive of the tangent to a curve at a point A as:

- a line passing through A but not crossing the curve in a neighbourhood of A (the point of view used notably by Appollonius to determine the tangents to the conics and not requiring a differential approach),

- a line having a double intersection with the curve at A (a point of view present in the works of Euler and Cramer for example, then later systematized in the context of algebraic geometry),

- a line passing through two points infinitely close to A on the curve (the point of view of Fermat, Leibniz,...) or the line which the curve becomes when one magnifies it in a neighborhood of A,

- the limit of the secants (AM) as the point M tends toward A along the curve, as in figure 18 (the viewpoint of D'Alembert for example, traditional in education),

- the best linear approximation or the only linear approximation of the first order to the curve in a neighborhood of A (leading to the more sophisticated idea of the tangent linear map),

Figure 18 : The dynamic movement of a secant to a tangential position

- the line passing through *A* whose slope is given by the derivative at *A* of the function associated with the curve (where the derivative is assumed to exist).

Similarly, one can see in the derivative at *x=a* of the function f as:

- the limit of the ratio (f(*x+h*)–f(*x*))/*h* when *h* tends toward 0,
- the first order coefficient of the expansion limited to order 1 of the function at *a* (as in the contemporary French programme),
- the coefficient of the first order term in the full series expansion of f around *a* (point of view of Lagrange),
- the coefficient characterizing the linear map, tangent to f at *a*,
- the slope of the tangent at *a*,
- the number or the function obtained by applying the usual rules of differentiation, knowing the derivatives of the elementary functions

or again,

- the slope of a highly magnified portion of the graph itself (for a "locally straight" graph – the viewpoint advocated by Tall (1986a) and now adopted by the British School Mathematics Project in its new curriculum).

Similarly the integral has several different conceptions: the inverse operation of differentiation, a process for obtaining lengths, areas, volumes, a continuous linear form on a space of functions, or more generally a process of measure.

One can imagine that these points of view can coexist in mathematicians, some of them being preferred because of the mathematical context or because of individual preference. One or another situation can lead to their being called forth and put into effect. What happens for students? In particular, which viewpoints are preferred and which are difficult to put into effect? Are there subtle transitions between different levels of functioning, are there conflicts, obstacles? What role does education play in all this?

Whilst the research discussed in the next section was not designed to answer these specific questions, it furnishes a good indication of the conceptions of students and the manner in which they develop.

2.1 A CROSS-SECTIONAL STUDY OF THE UNDERSTANDING OF ELEMENTARY CALCULUS IN ADOLESCENTS AND YOUNG ADULTS

The study conducted by Orton in his thesis (1980) is experimentally based on individual interviews conducted with 110 students aged 16 to 22 (60 in their last year of secondary school – the English "sixth form" – and 50 in college) all having chosen to study mathematics and having taken at least one course in calculus. The different tasks proposed in the interview were regrouped by items in the analysis, each concerning just one aspect of elementary calculus, and the responses were evaluated for each item on a scale of from 0 to 4. Tables V and VI, below – extracted from Orton (1983a, 1983b) – give a global idea of the levels of success corresponding to each item for the two populations involved.

Following the work of Donaldson (1963), Orton classified the errors into three categories: "structural errors", "executive errors", and "arbitrary errors".

Structural errors were those which arose from some failure to appreciate the relationships involved in the problem or to grasp some principle essential to solution. Arbitrary errors were said to be those in which the subject behaved arbitrarily and failed to take account of the constraints laid down in what was given. Executive errors were those which involved failure to carry out manipulations, though the principles involved may have been understood.

Without going into further details, it seems important to note that the research showed:

- *a reasonable mastery of algorithmic algebra* in terms of calculation of derivatives and primitives, at least for the simple functions, as indicated by the degree of success in tables V and VI.

- *significant difficulty in conceptualizing the limit processes underlying the notions of derivative and integral*: For instance, when questioned what happens in figure 19 to the secants PQ on a sketched curve as the point Q_n tends towards P on the circle, 43 students seemed incapable, even when strongly prompted, to

Description of task	Mean Scores (out of 4)	
	School	College
Infinite geometric sequences	2.88	2.56
Limits of geometric sequences	2.92	2.78
Substitution and increases from equations	3.32	3.68
Rate of change from straight line graph	2.22	2.02
Rate, average rate, instantaneous rate	0.88	1.18
Average rate of change from curve	2.22	2.02
Carrying out differentiation	3.62	3.50
Differentiation as a limit	1.88	1.14
Use of d-symbolism	1.52	1.40
Significance of rate of change from differentiation	3.43	3.62
Gradient of tangent to curve by differentiation	3.63	3.76
Stationary points on a graph	2.30	2.54

Table V : Performance on calculus tasks

Description of task	Mean Scores (out of 4)	
	School	College
Limits of sequences of numbers	3.28	3.06
Limits from general terms	2.82	2.90
Heights of rectangles under graphs	2.68	3.42
Use of previous heights in a new situation	2.40	3.12
Calculation of areas of rectangles	3.03	3.62
Simplification of sum of areas of rectangles	2.43	3.52
Sequence of approximations to area under graph	2.18	3.22
Limit of sequence equals area under graph	0.78	1.00
Limit from sequence of fractions, from general term	1.67	2.48
Carrying out integration	2.98	3.40
Integral of sum equals sum of integrals	1.10	0.60
Complications in area calculations	2.55	2.78
Volume of revolution	0.95	0.88

Table VI

see that the process led to the tangent to the curve?

There appeared to be a considerable confusion in that the secant was ignored by many students, they appeared only to focus their attention on the chord PQ, despite the fact that the diagram and explanation were intended to try to insure that this did not happen... Typical unsatisfactory responses included : "the line gets shorter", "it becomes a point", "the area gets smaller"...

These responses are entirely consistent with the nature of the geometric obstacle studied separately by Cornu and Sierpińska (see Chapter 10).

Similarly, although students showed certain competencies in calculating limits of sequences given explicitly in terms of numbers or simple functional expressions, only 10 were capable of expressing that the exact area under part of a parabola could be obtained as the limit of the sums of approximating rectangular strips. Orton concluded:

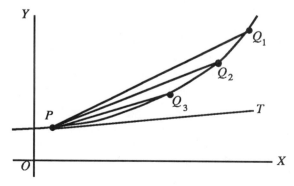

Figure 19 : Secants "tending to" a tangent

Students were able to obtain a limit from a sequence when the sequence was directly requested but were not able to appreciate when a limit would solve a problem.

A similar difficulty arises evaluating areas bounded by curves in slightly more general circumstances (the presence of negative values, discontinuities, or curves associated with functions $x=f(y)$ for example):

Many students appeared to know what to do, but when questioned about their method, didn't really know why they were doing it.

Further difficulties included:

- *the difficulty of using relevant graphical representations.* Students could usually calculate derivatives of polynomials correctly and were equally successful with a task in the form:

  Find the gradient of the tangent to the curve $y=x^3-3x^2+4$ when $x = 3$.

  But having to evaluate these same rates of growth from the graphs for functions of similar complexity, a non-negligible proportion made errors, confusing average and instantaneous rate of growth or simply giving the value of the function at the point in question. In a graphical context the expression of the derivative as a limit was poorly understood: 96 students, after having found the expression for average rate of growth of the function $f(x)=3x^2+1$, between a and $a+h$, could not see how to obtain the rate of growth at 2.5 or at a general x.

- *the minimal meaning ascribed to the symbols used.*

  For instance, when asked to explain the meaning of dx, dy, dy/dx, 71 gave incorrect responses for the rate of growth:

  "$\dfrac{\text{rate of change of } y}{\text{rate of change of } x}$", "rate of change at a point", "small increase in the rate of change"

  and 25 interpreted dx as the limit of δx when δx tends toward 0.

This strength in the algorithms of algebra as opposed to weakness in graphs and geometry is also found by other authors, some already cited in the preceding chapter, for example, Tall, 1977, Artigue & Szwed, 1983. The latter presents an account of the responses of 89 first year university mathematics students of mathematics to the question in figure 20.

The first question yielded 28 correct responses (with some minor errors in calculation), 29 incorrect responses, and 31 incomplete responses, the errors being principally based on confusion between continuity and differentiability or between differentiability and existence of derivatives to the left and to the right.

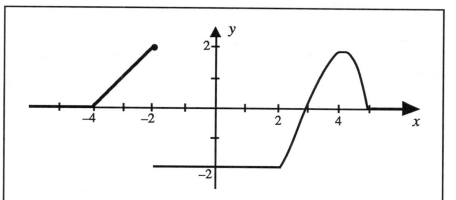

Let f be the function whose graph is drawn above

• at which points is f differentiable? differentiable to the right? to the left?

• describe the behavior of the graph of f

• describe the behavior of the graph of the function g defined by

$$g(x) = \int_{-4}^{x} f(t) \ dt.$$

Figure 20 : A conceptual task on differentiability and integration

For the graph of f':

• 67 students (out of 89) gave a correct graph on the portion]-4,2[and 5 others had an incorrectly positioned horizontal segment on this portion,

• 63 considered the sign of the derivative on the interval]2,5[, but only 18 students gave an acceptable form for the graph and, when the values attributed to f'(3) and f'(5) are taken into account, only 10 pupils gave a satisfactory graph.

The graph of the function g was only attempted by 35 students. The curves produced were extremely diverse and seemed to have only one property in common: the graph is a line segment on [-2,2]. Only 14 gave the correct slope and at least 14 graphs were discontinuous. Of the 35 graphs produced, only 13 considered the direction of variation in g, and only 3 could be considered acceptable solutions.

Analysis of the transcript and the errors committed shows that many students did not work directly with the graphs but sought to obtain algebraic expressions for f on each interval in order to differentiate or integrate them, in the latter case making numerous errors.

When the results produced were inconsistent with the graph, there seemed little awareness of conflict, their confidence being placed more in the calculation than the picture.

A simplified version of the derivative part of this test has been used for several years in university entrance examinations for potential mathematics students. The results obtained are in the same spirit: significant confusion over continuity and differentiability of the graphs, with correct responses only for linear parts of functions (cf. Robert, 1983; Authier, 1986).

2.2 A STUDY OF STUDENT CONCEPTIONS OF THE DIFFERENTIAL, AND OF THE PROCESSES OF DIFFERENTIATION AND INTEGRATION

To investigate students conceptions in the related disciplines of mathematics and physics, two teams from mathematics education and one from physics education collaborated in a research study of the effects of current educational practices in the first two years of university (cf. Alibert et al, 1987; Artigue & Viennot, 1987; Artigue et al, 1989). The researchers conducted their work in three directions:

- analysis of the historical evolution of the concepts and how they were taught,

- analysis of student conceptions and, to a lesser extent, of the teachers, through approximately 10 questionnaires from a mathematical or physical viewpoint, together with individual interviews,

- experimentation with, and evaluation of, sequences of instruction.

The analysis of historical evolution suggested three directions for analysis of students' conceptions:

- the meaning and usefulness of differentials and differential procedures,

- approximation and rigour in reasoning,

- the role of differential elements.

We consider the role of each of these in turn.

2.2.1 THE MEANING AND USEFULNESS OF DIFFERENTIALS AND DIFFERENTIAL PROCEDURES

Mathematical questionnaires, completed by 85 third year university students, revealed an important difference between the declarative level (how the students described the concepts) and the procedural level (how they carried them out). At the declarative level, the tangent linear approximation differential dominated, conforming to the definition in the course. At the procedural level, the differential tended to lose its functional role and the status of approximation disappeared, to be replaced by algebraic algorithms using partial derivatives and Jacobian matrices.

A typical manifestation of this was revealed in the responses to the following two questions:

"If you had to explain what a differential is to a first year student:

1a) What definition would you give?

1b) What notations would you introduce?

1c) What examples would you use?

1d) What important points would you stress?"

and

Is the function $f:\mathbb{R}\to\mathbb{R}$ defined by $f(x,y) = 2x+4y+y^3 (\sqrt{1-\cos x}+y)$ differentiable at the point $(0,0)$? Justify your answer.

The responses for 1a) were dominated by the notion of tangent linear map; in the remainder of the question:

- 33 mentioned the relation between the differential and related notions (continuity, differentiability, existence of partial derivatives),

- 15 mentioned the idea of local approximation, the functional and linear aspect,

- only 11 mentioned the algorithmic procedures of calculation.

In the second case, a small minority (13%) recognized that the given function is already in the form of a linear expansion of order one and almost all, in spite of the complication caused by the vanishing of the square root at the point in question, went directly to the calculation of partial derivatives. Even among those who recognized the linear part, very few succeeded in proving that the remainder is of higher order.

The percentages of responses to other questions confirm these impressions: 86% responded to a complicated calculation of second partials of a composite function at the end of the questionnaire; less than 10% responded to questions requesting a justification of classical approximations for calculating a volume by cutting it into slices. Some students even complained that they thought the latter was off the syllabus.

The questionnaires formulated from a physical viewpoint show clearly that first and second year students do not understand the differential procedures. For example, they were provided with the beginning of a classical calculation of atmospheric pressure leading to the differential expression $dp = -\rho g\, dz$. (figure 21).

The responses, summarized in figure 21, show:

- the strong conviction of students that the atmosphere needed to be cut into infinitesimal slices (90%),

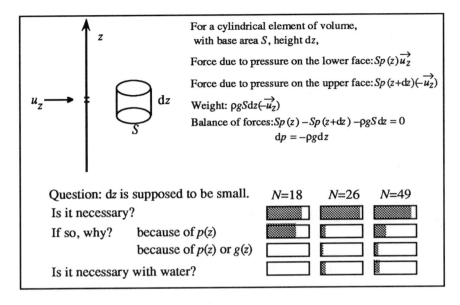

For a cylindrical element of volume,
  with base area S, height dz,

Force due to pressure on the lower face: $Sp(z)\vec{u_z}$

Force due to pressure on the upper face: $Sp(z+dz)(-\vec{u_z})$

Weight: $\rho g S dz(-\vec{u_z})$

Balance of forces: $Sp(z) - Sp(z+dz) - \rho g S dz = 0$

$$dp = -\rho g dz$$

Question: dz is supposed to be small. $N=18$ $N=26$ $N=49$

Is it necessary?

If so, why?   because of $p(z)$

              because of $p(z)$ or $g(z)$

Is it necessary with water?

Figure 21

but

• their justifications were based mainly on the mistaken view that p is a function
  of z, rather than the fact that ρg is not constant (it is a function of z).

With water the factor ρg is constant and cutting into slices is no longer necessary, but
students who give a wrong reason for "air" do not see this, still suggesting that it is necessary
to cut into slices.

2.2.2 APPROXIMATION AND RIGOUR IN REASONING

In general requests on the mathematics questionnaires for justification of approximations
were poorly answered and were considered "off the syllabus" unless they could be solved
by directly quoting classical theorems. In specific examples, the approximations suggested
by the students were incorrect more than half of the time and the great majority of errors
were concerned with the remainders, as if the fact of writing an ε arbitrarily at the end of
a formula is sufficient to make it rigorous.

In physics, the problem of rigour is handled differently, with phrases such as "provided
that dz is sufficiently small" being used instead of ε–δ methods.

These conceptions remained persistent, even with more advanced students. The
problem in figure 22 was presented to 50 third year mathematics students, 22 fourth year
physics students and 13 fourth year physics students in a very selective group preparing for
teaching.

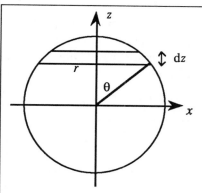

The volume of such a slice at height z is assimilated to that of a right cylinder of the same thickness dz and of base area: πr^2 (as shown in the drawing)

So $dV = \pi r^2 dz = \pi(R^2 - z^2) dz$

and the volume of the sphere is :

$$V = \int_{-R}^{+R} \pi(R^2 - z^2)\, dz = \frac{4}{3}\pi r^3$$

If the same procedure is used to find the area of a sphere, the following expression is obtained for the area of an elementary slice of thickness dz at height z:

$$dS = 2\pi r dz = 2\pi\sqrt{R^2 - z^2}\, dz$$

and therefore the area of the sphere is given by the integral:

$$S = \int_{-R}^{+R} 2\pi\sqrt{R^2 - z^2}\, dz = \int_0^{\pi} 2\pi R^2 \sin^2\theta\, d\theta = \pi^2 R^2.$$

Could you explain why the same method leads to a correct value in the first case (volume) and to a false value in the second one (area).

Figure 22

Only a small proportion of those questioned (less than 25%) seemed capable of giving an acceptable response. For instance, a future teacher wrote:

"Maybe it is just by chance that it works (computation of the volume). Indeed, the relationship $dV=S(z)\,dz$ is not true. It could be the reason why the other computation about the area of the sphere does not work."

The results obtained here are completely compatible with those obtained in another mathematical questionnaire given to 35 students of the third year (cf. figure 23).

The majority of students responding considered the different approximations to be valid because the given slices tend towards a spherical slice. One has the impression that, for them, this geometric convergence guarantees the convergence of all quantities associated with the figure, even though this is false.

To find the volume of a sphere, physicists cut it into elementary slices and approximate each slice by a small right cylinder. Why cylinders? Is it possible to choose other approximations, for instance, do the following approximations lead to the same result?

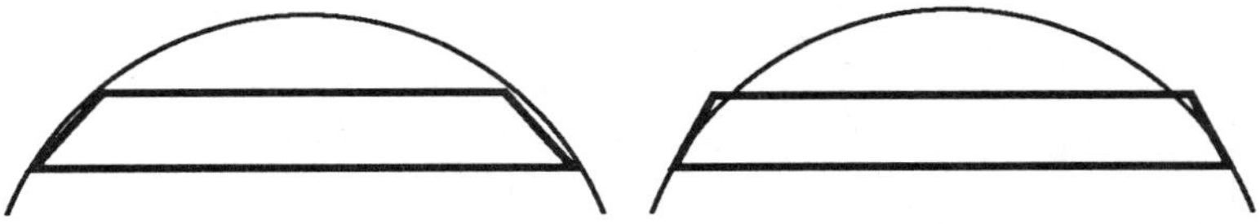

Slices generated by chords   Slices generated by tangents

Figure 23 : Elemental slices

2.2.3 THE ROLE OF DIFFERENTIAL ELEMENTS

The responses to the questionnaire show that the functional role of differentials is hardly present in mathematics students at the declarative level: less than a third of the differentials given in response to various questions were in the form of a function. Moreover, the geometric images associated with the concept were weak and restricted to one dimension. For instance, this is manifested in the responses to the following questions:

"The map f: $\mathbb{R}^2 \to \mathbb{R}$ is defined by:

$$f(x,y)=\exp(y^2+x)\cdot\sin(xy)$$

Find its differential at the point (1,0) and give a geometrical interpretation."

Of the 85 third year students questioned, only 31 dealt with the geometric interpretation and only 8 gave a correct interpretation in terms of the tangent plane. In particular, many spoke of the tangent to the curve as if they were still in the one dimensional case.

In physics, the results obtained show that the role of differential elements oscillates between two poles:

- At one extremity, the differential elements have a purely formal role of indicating the variable of integration and it was better to avoid thinking too much about what they could mean when manipulating them:

  "To integrate, it is essential not to think about what dl represents, but to proceed mechanically, otherwise we are done for" (student of the first year) – "dx is not real" – "immaterial" – "the length is fictitious" – "in fact, it does not matter at all, when integrating, dl becomes a variable of integration".

- At the other end, the differential elements have a very strong material existence which can exclude all other meaning:

  "dl is a small length", "a little bit of wire",

The problem is to find the magnetic field created by a current flowing in a wire of infinite length at a point M (as in the drawing below).

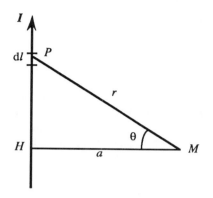

An element of wire dl, round a point P creates a field $\vec{dB}$ at M.

In this case, the vectorial addition leads to an algebraic one, with:

$$dB = \frac{\mu_0 I \, dl \cos\theta}{4\pi r^2}$$

Questions:

• can dB be considered as the differential of a function? If so, a function of what variables?

• same question for dl.

• express dB with only one variable.

Figure 24 : The role of differential elements

Between these two poles, there are a wide range of views such as,

"dz is the limit of Δz when $\Delta z \to 0$", "dl is an infinitely small element", "one cannot find anything smaller", "it means very simple" (which apparently means that differentials are often used to simplify the situation).

Often, to solve a problem, it is necessary to give a functional meaning to the differential elements. The authors consider that strong conceptions, either purely formal or purely material, can make this process more difficult. Thus, faced with the problem of calculating the induction created by a straight electric wire through which a current is passed (figure 24), the only students who did not make the classic error of mechanically transforming the differential element dl into a function of r, θ as $dl = r \, d\theta$, were exactly those who were also capable of giving a functional role to differential elements.

From the various results obtained, it seems that *two* differentials exists simultaneously for the students, potentially in conflict with the definitions, but governed by certain tendencies:

• the algebraic algorithms overwhelm the meaning and process of linear approximation,

- questions of rigour are reduced to formalism,

- the functional mode of thought is weak.

It should be emphasized that the existing gap between teaching in the two disciplines usually prevents the potential conflicts from being realized.

2.3 THE ROLE OF EDUCATION

The empirical results discussed so far show a broad coherence over a wide range of students: in every case there is a dominance of the algebraic mode of procedure contrasting with a frailty in geometric and graphic modes, and a lack of meaning for limits and/or approximation. Questions of rigour or justification linked to the treatment of approximations are conceived as secondary.

But how are these observations influenced by the educational process? Alibert *et al* (1987) reinterpret the results obtained in the context of their current instructional procedures in analysis. Two contexts appear: *algorithmic procedures* and *the conceptual viewpoint* involving questions of meaning and legitimacy of the theory. Each discipline tries to manage the relations between these two contexts, in a manner appropriate to the subject, whilst seeking an optimal balance between rigour and operational practice. The conceptions developed by students are essentially a reflection of the very unsatisfactory equilibria found by education.

In mathematics the means of justification is classically that of proof. However, from the start, education distorts real difficulties concerning limits, functions, basic tools of approximation (such as inequalities, absolute values, reasoning with sufficient conditions, &c), and the understanding and manipulation of quantified statements. Instead it conceals them all by using powerful algebraic algorithms (calculation of derivatives, partial derivatives, Jacobian matrices, primitives) and potent theorems which reduce theoretical considerations to algebraic techniques (such as theorems involving the sum, product and composition of C^1 functions). Regrettably this premature algebraic algorithmization, on the one hand gives too privileged a role to the algebraic setting, and on the other tends to drain the differential and integral procedures of their real meaning.

In physics, not being constrained to requiring proofs, it is possible to take refuge in the conviction that it works, even if one does not know why, once again denying students a satisfactory scientific experience.

3. RESEARCH IN DIDACTIC ENGINEERING

The work discussed in this section complements the research described in the preceding one by using acquired knowledge about the learning process, and the effects of the usual pedagogy, to develop and test new methods of teaching and learning. The French term *ingénierie didactique* for this activity translates literally as "didactic engineering".

3.1 "GRAPHIC CALCULUS"

"Graphic Calculus" is an approach to the calculus developed by Tall (1986a, 1986c, 1990) which acknowledges the known conceptual obstacles in the limit concept and proposes instead a new learning sequence built on the visualization of the "local straightness" of graphs. It uses the graphic and dynamic possibilities of the computer to give a cognitive base for the notions of derivative and integral in secondary education which can lead to later formalizations in either standard or non-standard analysis.

In order to achieve this objective, the student is furnished with a computer environment (or microworld) designed to encourage the exploration of examples of specific mathematical processes and concepts. This requires a special kind of software:

An environment that provides the user the facilities of manipulating examples (and, where possible, non-examples) of a concept, I term a *generic organizer*. The word "generic" means that the learner's attention is directed to certain aspects of the examples which embody the more abstract concept. Thus the equality 3+2=2+3 may be seen as a specific example of the commutative property of addition. The generic example is seen as a representative of the whole class of examples which embody the general property.

Tall recognizes that a generic organizer does not guarantee its use by the student as a tool of abstraction and suggests the need for an "organizing agent":

"guidance from a teacher, a textbook or appropriate computer material."

For this reason the instructional treatment is based on three phases:

- a first phase of familiarization and negotiation of meaning. It is conducted in the form of a dialogue between the teacher and the students, a dialogue designated as the "enhanced Socratic mode", the term "enhanced" referring to the aid provided by the computer to communication:

  "The mathematics is no longer just in the head of the teacher, or statically recorded in a book. It has an external representation on the computer as a dynamic process."

- a second phase of autonomous work by students with the generic organizer,

and finally,

- a last phase of discussion and evaluation looking towards establishing the point and making sure that the concept images constructed by the pupils are compatible with those of the community of mathematicians.

The generic organizers elaborated by the author are specifically adapted for introducing the ideas of derivatives, integrals and differential equations.

For instance, the software "Gradient" contains various modules permitting the graphing of functions defined by a formula, magnifying a graph around a point, superimposing onto the graph $y=f(x)$ the line passing through the points $(x,f(x))$ and $(x+c,f(x+c))$ (for fixed c),

f(x)=sinx

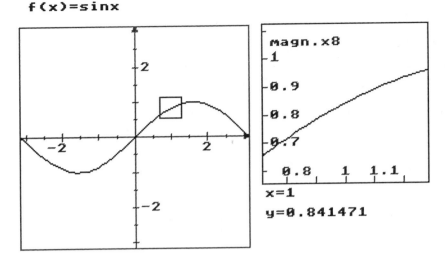

Figure 25 : a locally straight curve (magnified a little)

and dynamically seeing the evolution of this line and the curve representing its slope, as x varies. It also allows the superimposition of a graph, to test conjectures made about the derivative of this or that function (monomials or trigonometric functions for example). Finally, as an archetypal non-example, it contains a module which shows the construction by steps of a function continuous and non-differentiable at every point: "the blancmange function".

As conceived by the author, these generic organizers are designed to permit the pupil at the pre-calculus level to develop a concept image of the notion of derivative based on:

- The conception of a function differentiable at a point as a function which, by magnification around the point, eventually becomes like a straight line.

- A global image of a derived function associated with the notion of "practical tangent": a straight line passing through two points very close to each other on the curve.

Corresponding generic organizers have also been designed for integration and differential equations. These are discussed in greater detail in chapter 14.

Experimental studies using generic organizers described above were done in several classes of secondary education with other classes as controls. The results confirm on the one hand the difficulties already cited in §2 of conceptualization of the notion of a limit, both in terms of tangent as a limit of secants and of the derivative as a limit of slopes of secants. The results of the post-test differed little from those of the pre-test, in the experimental groups as well as the control groups. Tall concluded:

f(x)=bl(x)

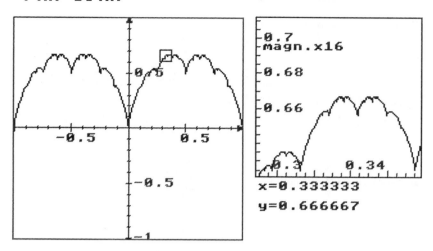

Figure 26 : a highly wrinkled function that is nowhere locally straight

"The low level of responses indicate the high cognitive demand of this general concept, reinforcing the opinion that, although the notion of a limit is the natural foundation of a mathematical development of the calculus, it is not a natural starting point for a cognitive development".

(Tall, 1986d)

As a result of the instruction, a significant number of the experimental pupils tended to conceive of the tangent as a straight line passing through two very close points on the curve. But they were far better at recognizing and drawing derivatives, even attaining performances comparable to those of university students. For example, 67% of the experimental students recognized and justified that graph (2) in figure 27 (overleaf) is the one whose derivative is graph (1), (with 68% for university students), while only 8% in the control group were successful.

Tall concludes in these terms:

"Once more empirical research has demonstrated a process of didactic inversion that gives an attractive cognitive approach. In this case, the cognitive approach, in the shape of the practical tangent, proves to be surprisingly good mathematics." (Tall, 1986d)

Other approaches using the computer to reorganize the syllabus and to introduce the calculus without using the limit as a prerequisite have shown success. D'Halluin & Poisson (1988), for instance, pursued research on "a strategy for teaching mathematics: the mathematization of situations integrating the computer as a tool and as a mode of thought" as continuing education for adults and school drop-outs. A function has three formal objects associated in interaction : Picture, Graph, Formula (the triple "PGF") on which the computer permits global operations. The introduction of the differentiation-integration concept takes place first at the numerical and graphical level, exploiting the computer, starting from a physical situation: the construction of a road. The problems of pitch and of

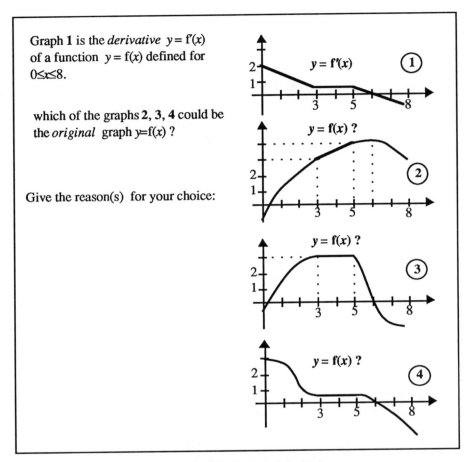

Graph **1** is the *derivative* $y = f'(x)$ of a function $y = f(x)$ defined for $0 \leq x \leq 8$.

which of the graphs **2, 3, 4** could be the *original* graph $y = f(x)$?

Give the reason(s) for your choice:

Figure 27 : Recognizing an anti-derivative

digging and filling give a meaning to the derivative corresponding to that of the practical tangent of Tall and to the integral in terms of measure. From the beginning they emphasize reciprocity of the operations of differentiation and integration through calculation of difference tables for the slope, tables of sums for the areas, global visualization of the slope and area curves. They next ask the students to study situations of speed (motion approach) and of distribution of salaries (statistical approach to the integral). The algebraic operationalization comes later, building on simple calculations of slopes and areas, using previously developed tools.

3.2 TEACHING INTEGRATION THROUGH SCIENTIFIC DEBATE

In recent years, a form of scientific debate has been introduced at the University of Grenoble (Legrand *et al*, 1986). Full details of the methodology will be discussed in the chapter 13 on proof in advanced mathematical thinking. Here we will concentrate on the outcomes of this approach in the learning of concepts in analysis. It occurred in a context where students were encouraged to conjecture and debate ideas in groups within a large class, where arguments were proposed and addressed to other students rather than the teacher.

The concept of integral was studied in great depth during this research. The associated didactic engineering was presented to first year university students who had already studied the secondary curriculum described earlier, including calculation of simple primitives and the conception of the integral both as the inverse operation of differentiation and the area under a curve. The new curriculum was designed to enrich the conceptions of students by giving a meaning to the notion of integral procedure.

It began with the following problem (Alibert *et al*, 1987b):

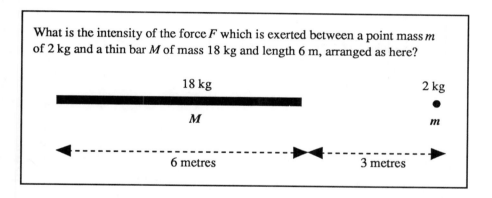

What is the intensity of the force F which is exerted between a point mass m of 2 kg and a thin bar M of mass 18 kg and length 6 m, arranged as here?

18 kg   2 kg

M m

6 metres   3 metres

figure 28 : A problem to initiate debate in analysis

The researchers hypothesize that, appropriately managed, scientific debate can help to solve this problem. One method is through visualizing the bar as being made up of tiny slices, calculating the force, then refining and passing to the limit through integration. However, the vast majority of students suggest a solution by conceiving the mass of the rod concentrated at the centre of gravity, which proves erroneous. In the course of the experiments, students were always found who proposed testing the validity of such a calculation by cutting the bar in two and applying the principle of centre of gravity to each half of the bar, which gives a different answer, so the principle is seen to be in conflict with itself. If the idea of concentrating the mass at a point is to be retained, new methods are suggested by placing the point mass at one end, or the other, to obtain inequalities. Repeating such process on each half of the rod, then on each quarter, and so on, leads to a conviction that one is going to be able to obtain a value to any desired precision, prior to the passage to the limit.

This first problem was followed by others aimed at extracting the integral procedure from other situations (averages, problems using time as a variable, space variables in two or three dimensions). Only after these problems is the study of mathematical properties of the integral operation begun, largely based on the work developed by the students.

The results of the first two experiments conducted in first year at Grenoble University with 105 students in 12 two hour sessions and 101 students in 14 two hour sessions are reported in Legrand et al, (1986). The effects of scientific debate within the overall instruction revealed improved understanding of integration in the final exam. For instance, the problem in figure 29 was given in 1986.

Let f be the function defined on the interval $\Omega = [0,4]$ of $\mathbb{R}$ as follows:

$$\forall x \in [0,2], \quad f(x) = \int_0^x e^{-t^2} dt$$

$$\forall x \in]2,3[, \quad f(x) = 2$$

$$\forall x \in]3,4], \quad f(x) = \frac{1}{4}\sqrt{91-x^3}$$

and $f(3) = 1$.

Give a rough sketch of the graph of f on Ω.

Determine the subset U of solutions in Ω of the equation $f(x) = \frac{3}{2}$.

Let $F(x) = \int_2^x f(u)\ du$

(a) draw a sketch of the graph of F on its domain of definition,

(b) show that the equation $F(x)=2.5$ admits a unique solution α on Ω,

(c) write a Pascal program which returns a value x in Ω
    for which one is sure that $|F(x)-2.5| \le 10^{-2}$.

CONJECTURES:

Resolve the following conjectures

Conjecture 1: If f is an integrable function on a simple domain
    and satisfies $\int_\Omega f = 1$,
    then for each positive constant k one has $\int_\Omega (f+k) = 1+k$.

Conjecture 2: If U is the rectangle $\{(x,y) \in \mathbb{R}^2 : 0 \le x \le 2, 0 \le y \le 1\}$
    and $g(x,y) = e^{2x-y^3}$, then $\int_U g > 2$.

Figure 29

89 students (84%) passed this test and only 20% tried to solve the problem by looking for primitives. This led the authors to conclude:

"The results show that a majority of students acquired a satisfactory level of understanding of the integral concept introduced by the method of scientific debate and that they understand it in sufficient depth as to know how to explore their knowledge, even when the usual algorithms are not applicable."   (Alibert *et al*, 1987b)

3.3 DIDACTIC ENGINEERING IN TEACHING DIFFERENTIAL EQUATIONS

Traditionally differential equations have been taught as a catalogue of recipes for algebraic solution in the classical integrable cases. Recently software has been developed for the effective teaching of differential equations by a number of authors (e.g. Tall, 1986b; Tall *et al*, 1990; Koçak, 1986; Hubbard & West, 1990). The research reported here, conducted by Artigue (1987), is aimed at studying the viability of a teaching approach which tries from the beginning to coordinate the algebraic, numerical, and graphic approaches with the solution of associated differential equations. The author interpreted present-day instruction as a position of stable equilibrium of a system subject to a set of constraints (epistemological, cognitive, didactic conventions, the available mental representations of the students and teachers) and seeks to modify some of these to permit the system to come to another stable equilibrium that is more satisfying from the point of view of the epistemology of the field. Computer software constitutes the principal lever employed to modify the space of constraints, facilitating access to numerical and graphical representations. In the new curriculum it is used in both interactive and ready prepared form (using supplied computer-generated graphs) to take account constraints of time and material, and to optimize management.

For example, in the graphical context, it is first used in prepared form to give meaning to the qualitative solution: drawing curves compatible with a field of tangent directions. It is then used to introduce elementary qualitative tools: isoclinic lines, in particular, the isoclinic line 0 as being essential to determine the direction of variations of the solutions that act as barriers or separatrices between different types of solutions, to identify stability by simple geometric transformations (such as symmetry, translation). Such an approach proves to be motivating in significantly improving students' abilities to associate pictures of solutions with algebraic equations in a way which is far less complex than having to make drawings by hand (figure 30).

Matching equations and drawings (8 equation and 8 drawings were given in the original test). Here are some the equations and two of the drawings.

$$y' = \frac{y}{x^2-1}, \ y' = y^2 - 1, \ y' = 2x + y, \ y' = \sin(xy),$$

$$y' = \frac{\sin(3x)}{1-x^2}, \ y' = \sin x \sin y, \ y' = y + 1.$$

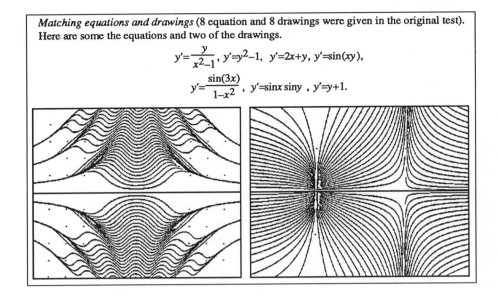

Figure 30 : Relating graphical and symbolic representations

The computer is also used in an interactive manner. For instance, when studying differential equations depending on a parameter; the software allows experimental determination of the different types of pictures of possible phase portraits, then students have to justify the most likely graphs obtained by using algebraic and/or graphical arguments, to address a list of unsolved problems and to formulate, if possible, appropriate conjectures.

Although only a short time was available for the instruction, research revealed very positive results. The students showed themselves capable of giving meaning to the qualitative approach, to describe and draw solutions without algebraic integration in simple cases such as the equation:

$$y' = \left(\frac{1}{1+x^2}\right)^2 - y^2$$

and to coordinate the algebraic and graphical contexts.

3.4 SUMMARY

It is important to emphasize once more the convergence of the research. In the work on didactic engineering there are themes which relate the different approaches:

- focus on construction and the control of meaning,

- search for a better equilibrium between the different representations for the concepts, in particular, concern with a better use of a graphical context,

- concern to use the possibilities offered by the computer to rethink the content of education in terms of epistemological adequacy of the domains considered and the cognitive capacity of students.

As for the precise contents envisaged by the research, even where the works concern different levels, one finds again common preoccupations:

- concern with developing a functional approach,

- concern to focus the notion of derivative on the existence of a good approximation of the first order, the computer allowing exact visualization of this property by magnification of the graph, even before the notion of limit is mastered.

However, the research is largely focussed on the intuitive beginnings of the subject and it is natural to ask the question: can the development of strong conceptions of this type subsequently form an obstacle to the construction of more formal concepts such as the measure interpretation of the integral procedure? On this point there is need for more empirical research.

The differences between these various experiments are more concerned with the proposed management of instruction. In all didactic engineering there is a place for an experimental approach to mathematics and organization of the construction of knowledge around the activity of the pupils. But the researchers at Grenoble also have the conviction that in order to allow the students to establish a correct epistemology of mathematical knowledge, it is necessary to change their relationship with mathematics created during their schooling, which is based on the predominance of an algorithmic approach and on a vision of proof as a simple contractual agreement rather than a means of convincing or of lifting uncertainty. In association with this, there is a conviction that this change can only be made by a break with the traditional learning contract through the notion of scientific debate. Such a conviction is developing in a number of different countries (cf. for example Schoenfeld, 1985 or Robert in chapter 8) and researchers seek to exploit the capacities of pupils at the end of secondary education to help them reflect explicitly on the mathematical processes.

4. CONCLUSION AND FUTURE PERSPECTIVES IN EDUCATION

The results obtained in the various research projects described in this chapter, present a strong coherence not only between themselves but also with those of the preceding chapter on pre-calculus concepts. Learning the beginnings of analysis presents certain difficulties which appear to be due to different factors, in particular:

- the highly sophisticated level of structure of the objects in the foundations of this conceptual field, such as sequences and functions,

- the existence of various obstacles, including those evident in the historical development, those due to the conflicting everyday meaning of some of the terms, obstacles due to the all-pervading problem of infinity, and to pupils' conceptualizations of the reals more consonant with the non-standard theory than the standard formalization,

- difficulties posed in learning specific techniques of the field: such as use of upper and lower bounds, use of the completeness axiom, reasoning by sufficient condition which forces the acceptance of loss of information,

- finally the difficulties due to formalization in this field: first because it introduces structural definitions which may conflict in the students mind with more intuitive spontaneous conceptions, and secondly because it bases proofs on complex propositions involving quantifications which operate in a direction seemingly contrary to the dynamic flow of intuitive thought.

These difficulties are far from being resolved by students in secondary education and are reflected in learning calculus which constitutes the most important part of education in mathematics during the first two years of the university. The research suggests that, faced with these difficulties, the usual instruction takes refuge in an intensive "algebraisation" of analysis: manipulating formulae rather than functions, emphasizing the calculation of derivatives rather than the theory of linear approximations, calculating primitives in integration rather than delving into the meaning of the integration procedure and learning recipes for solving differential equations without developing a general numerical or graphical approach to the solution. Moreover, one tries to resolve difficulties due to formalization by first giving definitions then quickly proving or quoting powerful theorems which permit the learner to move on from the subtle theory to return to algebraic algorithms.

The results of the research cited bring out clearly the perverse effects of this avoidance: in avoiding difficulties of formalization and techniques of approximation, a real chasm is created between concept definition and concept image; in excessively emphasizing the algebraic approach through facile algorithmization, the possibilities of changing points of view, essential to the real practice of the mathematician, are reduced. Furthermore it excludes the possibility of taking advantage of cognitive diversities which can exist among students. By avoiding the problems of legitimization one leads students to see proofs as a simple matter of didactic contract – something to be done to fulfil the requirements of the course. Certainly the students come to obtain a reasonable level of success in a certain number of algorithmic tasks, but it must be emphasized that there is not, in the context of this instruction, a real introduction to analysis: the conceptions developed by the students are poor and the subtle techniques in the field are not adopted.

However, the different experiences of didactic engineering presented here prevent fatalism. They tend to show in particular that the cognitive capacities of students, properly exploited, could warrant a more satisfying equilibrium between conceptualization and algorithmization as well as between the different contexts in which the concepts arise. And, as Dubinsky and Tall will discuss in chapter 14, use of appropriate computer languages can help the students handle quantifiers and construct a meaningful concept image of rigorous proof. Thus we see that the computer offers a number of didactic advantages:

- it provides possibilities for dynamic visualization to make the geometric and graphical contexts much more accessible and, properly exploited, it can help to bring out the necessary relations between algebraic and geometric representations,

- if the graphical context becomes more familiar, the unity of the graphic representation of the functional object which it furnishes can help to establish the concept image of the foundational concepts by enriching the stock of mental images,

- through experimental activities with the interactive simulations, students may be initiated into mathematics as a constructive scientific activity.

- appropriate computer languages can help with the problems of formalization, the constructive use of quantifiers and the development of rigour (as we shall see in chapter 14).

However, the computer is not a tool that will miraculously solve all the problems of teaching analysis. It can without doubt help overcome certain difficulties, but the different studies reported in this chapter show clearly that it will only be effective within a coherent teaching/learning context. Elaboration, experimentation and evaluation of such an approach is costly work. Moreover, even if it can help solve certain problems, the introduction of the computer tool into education cannot fail to create, in its turn, new problems in classroom management (availability of computers, coordination of computer use with other supports for learning, and so on) and even cognitive problems which, in their turn, should be the subject of further research.

We mentioned non-standard analysis in the first part of the chapter and at this point we return for further consideration.

The results of the research presented in the preceding chapter show that students' mental representations of the reals seems closer to non-standard representations then to standard representations. Some of them are not in perfect agreement with classical non-standard representations: for example when students said that 0.999... is the last number before 1, they adopt an atomistic point of view incompatible with the axioms of non-standard analysis, but one could imagine that it would be easier to move these conceptions towards coherent non-standard conceptions, which might then lead back into standard analysis.

The results of the research show equally the importance of the void that separates concept definition and concept image for school and university students and the operational deficiency of given definitions. Non-standard definitions are closer to the descriptions of differential and integral problems in physics than standard analysis. They also have fewer quantifiers and do not require the reversing of direction of the standard ε–δ or ε–N formulations: for example, a sequence u_n is convergent to a limit 1 if and only if for every infinitely large N, u_N–1 is an infinitesimal. Perhaps the definitions are more useable by students and the chasm between concept image and concept definition may be diminished by permitting a more gradual initiation to formalization.

The few pieces of research to date suggest that the logical baggage for current simplified introductions does not constitute a severe obstacle for students (cf. for example, Tall, 1980b, Artigue *et al*, 1985). But at the present time we cannot say what difficulties will be introduced by a non-standard approach to analysis, no more than we can say how standard and non-standard concepts might be coordinated in the minds of the students, nor what problems this coordination could pose. At the very least, however, the non-standard approach seems an interesting road to pursue in future research, if institutional conditions permit.

CHAPTER 12

THE ROLE OF STUDENTS' INTUITIONS OF INFINITY IN TEACHING THE CANTORIAN THEORY

DINA TIROSH

Here a difficulty presents itself which appears to me insoluble. Since it is clear that we may have one line segment longer than another, each containing an infinite number of points, we are forced to admit that, within one and the same class, we may have something greater than infinity, because the infinity of points in the long line segment is greater than the infinity of points in the short line segment. This assigning to an infinite quantity a value greater than infinity is quite beyond my comprehension.   (Galileo, 1638)

Infinity is undeniably one of the central concepts in philosophy, science and mathematics. In this chapter we review the nature of this concept and find that in different contexts the term infinity means different things, it might be *potential infinity* (representing a process that is finite and yet could go on for as long as is desired), or *actual infinity* in the sense of the cardinal infinity of Cantor, or *ordinal infinity*, also in the sense of Cantor, but this time representing correspondences between ordered sets, or *non-standard infinity* which arises in the study of non-standard analysis, and, unlike the others, admits all the operations of arithmetic, including division to give *infinitesimals*. It is clear that with this wide variety of technical meanings which often have quite different, even conflicting, properties, the possible intuitive meanings that arise in various contexts are likely also to be varied and in conflict. Indeed this is a common fact to be found throughout the research on the cognitive nature of the concept images associated with infinity. They are usually transient, unstable and conflicting. In this chapter we will first review the different perceptions of infinity in which we shall see that experiences of everyday life give little preparation for the nature of the cardinal infinity encountered in set theory.

   Given the conflict between previous experience and the formal theory, this is therefore an ideal opportunity to test the theories enunciated in previous chapters in which students are confronted with the cognitive obstacles and encouraged to reflect on them in an effort to re-construct their knowledge to come to a new and richer cognitive equilibrium. The second part of the chapter lays the groundwork by reporting a sustained investigation into the intuitive criteria that students use to determine whether two infinite sets have the same number of elements. The final part of the chapter considers a research study to help students develop a formal knowledge of the Cantorian set theory supported by an adequate intuitive background. It offers a possible model for the way in which students may be helped to come to terms with the kind of abstract thinking that causes such difficulty in the transition to advanced mathematical thinking. In this case the students are in the later years of secondary school and are therefore at a suitable stage of development to test out value of reflection on cognitive obstacles to assist transition to the more abstract forms of thinking at higher levels.

1. THEORETICAL CONCEPTIONS OF INFINITY

In the early history of mathematical development the two competing ideas of infinity were *potential infinity* in which a mathematical process can be carried out for as long as required to approach a desired objective, and *actual infinity* in which one contemplates the totality of infinity, through, for example, conceiving the totality of *all* natural numbers at one time.

Ever since Aristotle, philosophers and mathematicians have invariably rejected the concept of actual infinity. Aristotle himself argued that

"The infinite is potential, never actual"   (Aristotle, Physics, Book 3, Ch. 7).

Similarly, in 1831 Gauss stated

"I protest above all the use of an infinite quantity as a completed one, which in mathematics is never allowed. The infinite is only a *façon de parler* in which one properly speaks of limits".

(Gauss, in Dauben, 1983).

Rejection of the notion of actual infinity can be found even at the beginning of the twentieth century: Poincaré, in an essay on "The logic of infinity", noted that

There is no actual infinity, and when we speak of an infinite collection, we understand a collection to which we can add new elements unceasingly.   (Poincaré, 1963/1913, p. 47).

Why have mathematicians argued so consistently against actual infinity? A main source of their opposition to the idea is that it has given rise to numerous paradoxes and difficulties in mathematics. Even those mathematicians who essentially accepted the existence of actual infinity, such as Galileo, Bolzano, Dedekind, Hahn, Hilbert and Russell, were aware of the difficulties involved. Galileo, for example, pointed out that if the number of natural numbers is not only potentially but actually infinite, then there are as many perfect squares as there are natural numbers, since for every natural number there is a perfect square and every perfect square has a square root. He further noted that it is also possible to determine, on the basis of the "part-whole" principle, that there are more natural numbers than square numbers. Galileo concluded that infinite quantities are incomparable. He was one of the first to mention that

Difficulties arise when we attempt, with our finite minds, to discuss the infinite, assigning to it those properties which we give to the finite and limited.   (Galileo, 1954/1638, p. 31).

Cantor made a significant and surprising breakthrough in creating a theory of actual infinity. He defined not one, but two, distinct kinds of infinite numbers: transfinite *ordinal* numbers, which are denoted by ω, $\omega+1$, $\omega+2$,... 2ω, etc.; and transfinite *cardinal* numbers which are denoted by $\aleph_0$, $\aleph_1$, $\aleph_2$, etc. The transfinite ordinal numbers, which were the first to be introduced, are an extension of the notion of ordinal numbers to the infinite case. They were defined only for ordered sets. Two ordered sets are considered to have the same ordinal number if they can be put into a 1-1 correspondence with one another in such a manner as to maintain the order relation between corresponding elements. The transfinite cardinal numbers are an extension of the notion of counting. Two sets are considered to have the same cardinal number if they can be put into a 1-1 correspondence with each other. These two notions of transfinite numbers are clearly distinct from one another. In fact, Cantor showed that it is possible to construct an infinite number of infinite sets having different ordinal numbers but the same cardinal number.

Cantor's treatment of infinite sets leads to unexpected conclusions, such as: there are as many odd numbers as natural numbers; the number of even numbers is equal to the number of rational numbers; and the number of points in a line segment is greater than the number of natural numbers. These properties of infinite numbers are so startling that not a few of Cantor's contemporary mathematicians and philosophers were reluctant to accept the new doctrine. Even Cantor himself admitted that certain conclusions deriving from it appeared to be counter-intuitive.

2. STUDENTS' CONCEPTIONS OF INFINITY

The Cantorian set theory is the most commonly used theory of infinity today. Yet, recent psycho-didactical studies have shown that students face great difficulties in acquiring various properties of cardinal infinity that give the impression of being impossible or even self-contradictory (Fischbein, Tirosh, & Hess, 1979; Fischbein, Tirosh & Melamed, 1981; Duval, 1983; Borasi, 1984; Borasi, 1985; Tirosh, 1985; Martin & Wheeler, 1987; Wheeler & Martin, 1988; Tall, in press). It has been found that:

1. There are profound contradictions between the concept of actual infinity and our intellectual schemes, which are naturally adapted to finite objects and finite events. Consequently, some of the properties of cardinal infinity, such as the fact that $\aleph_0+1=\aleph_0$ and $2\aleph_0=\aleph_0$ are very difficult for many of us to swallow (Fischbein, Tirosh, & Hess, 1979; Fischbein, Tirosh & Melamed, 1981; Tall, 1980c, 1981, in press; Duval, 1983).

2. Intuitions of actual infinity are very resistant to the effects of age and of school-based instruction (Fischbein, Tirosh & Hess, 1979; Martin & Wheeler, 1987; Wheeler & Martin, 1988). This means that what we consider as self-evident concerning the magnitude of infinite sets remains largely unchanged from the age of 12 on, and these intuitions are unaffected by regular mathematical training which strengthens the logical schemes which are genuinely finitist.

3. Intuitions of actual infinity are very sensitive to the conceptual and figural context of the problem posed (Fischbein, Tirosh & Hess, 1979; Martin & Wheeler, 1987).

4. Students possess different ideas of infinity which largely influence their ability

   to cope with problems that deal with actual infinity (Sierpińska, 1987, 1989). These ideas are usually based on the notion of potential infinity (Fischbein, Tirosh & Hess, 1979).

5. The experiences that children encounter with actual infinity rarely relate to the notion of transfinite cardinal numbers. But they do have increasing experiences in school of quantities which grow large or small (Tall, 1980c, 1981).

Based on this observation, Tall formulated another notion of actual infinity (1981), *infinite measuring numbers*, which generalize the notion of measuring from real numbers to a larger number system. This corresponds in formal mathematics to an extension of the field of real numbers, such as the hyperreal number system of non-standard analysis. In this notion of infinity a line segment twice as long as another line segment contains twice as many infinitesimally small points as the other, and a line contains more points than a line segment. Tall argues that experiences of infinity that children encounter are more related to the notion of infinite measuring number and are closer to the modern theory of non-standard analysis than to cardinal number theory. For instance, in Tall, 1980d, he asked students to compute various limits, including the limits of

$$\frac{n^2}{n^2+1} \text{ and } \frac{n^5}{(1.1)^n}$$

as n tends to infinity. A student who wrote

$$\frac{n^2}{n^2+1} \rightarrow \frac{\infty}{\infty} = 1$$

was shown that a similar argument would give

$$\frac{n^5}{(1.1)^n} \rightarrow \frac{\infty}{\infty} = 1$$

but she replied firmly "no it wouldn't, because in this case the denominator is a *bigger* infinity, and the result would be zero". In this case Tall claims that her intuition is based more on extending experiences of comparative size that on potential infinity, and therefore is more akin to measuring infinity.

Fischbein *et al* (1979) cite an example where

$$1+\frac{1}{2}+\frac{1}{4}+\frac{1}{8}+\ldots$$

is stated to be

$$s = 2 - \frac{1}{\infty}, \text{ "because there is no end to the sum of segments".}$$

Here the potential infinity of the limiting process leads to a limit concept where the student divides by an infinitely large number to get an infinitely small one. This too is a closer fit with non-standard analysis than with cardinal numbers where infinities cannot be divided.

Following such cases Tall (to appear) suggests:

Most experiences with limits relate to things getting *large*, or *small*, or *close* to one another. All of these extrapolate experience from *arithmetic* rather than comparisons between sets and are more likely to evoke *measuring infinity*, rather that *cardinal infinity*. It follows that the ideas of *limits* and *infinity*, which are often considered together, *relate to two different and conflicting paradigms.*

These findings clearly indicate that our primary intuitions are not adapted to the notion of cardinal infinity. Thus, it would seem to require a considerable effort to develop appropriate "secondary intuitions" (i.e., intuitions which are acquired through educational intervention) of the notion of cardinal infinity. Such secondary intuitions are in conflict with some of our deeply held convictions, such as that the whole can not be equivalent to any of its parts and that there is only one level of infinity.

In the next section we consider the intuitive criteria adopted by students when determining whether two infinite sets have the same number of elements. In the section which follows we describe a research study in which students are assisted in constructing an adequate intuitive background for Cantorian set theory to lay the foundations for a formal knowledge of cardinal infinity.

2.1 STUDENTS' INTUITIVE CRITERIA
FOR COMPARING INFINITE QUANTITIES

Only a few studies have investigated students' intuitions concerning the comparison of infinite quantities (Duval, 1983; Fischbein, Tirosh & Hess, 1979; Martin & Wheeler, 1987; Sierpińska, 1989). In one of these studies, which is described fully in Tirosh (1985), 1381 students in the age range 11-17 years (grades 6-11) were given 32 mathematical problems that called for a comparison of infinite quantities. In each of these problems two infinite sets, with which the students were relatively familiar, were given. The students were asked to determine whether the two sets were equivalent and to justify their answers. A sample of the problems and the distribution of the students' answers appears in Table VII.

In line with the results of the above mentioned studies, it was found that students' responses to the problems included in Table 1, as well as to other problems which are not

The Sets compared A vs B		Same Cardinal	Different Cardinal	No Answer
The positive even numbers	The positive odd numbers	85*	15	0
All the points in a line	The natural numbers	80	19*	1
All points in a line segment	The natural numbers	56	42*	2
The natural numbers	The positive even numbers	48*	51	1
The natural numbers	The rational numbers	46*	51	3
All points in a line segment	All points in a line	40*	59	1
All points in a line segment	All points in a square	39*	59	2

* Correct answers

Table VII: Solutions to Problems Dealing with Equivalent Sets (%)

presented here, were relatively stable across the age groups. Moreover, students of various ages used the same intuitive criteria to compare two infinite sets.

The main argument used by the students to justify their claim that two sets have the same number of elements was: "All infinite sets have the same number of elements". For instance, 80% of the students claimed that there is an infinite number of natural numbers and also an infinite number of points in a line, and therefore there are as many natural numbers as points in a line.

The claim that two infinite sets were not equivalent was justified by one of three arguments:

(a) "A proper subset of a given set contains fewer elements than the set itself." For example, 51% of the students claimed that there are fewer positive even numbers than natural numbers, since the former is a proper subset of the latter;

(b) "A bounded set contains fewer elements than an unbounded set." For instance, 12% of the students used this argument to justify their claim that the number of the points in a square is greater than that in a line segment;

(c) "A linear set contains more elements than a two dimensional set." This argument was used by 38% of the students to justify the claim that there are more points in a square than in a line segment.

The following observations were made with respect to the students' responses.

(a) A very small percentage of the students (less than 1%) intuitively employed the notion of 1-1 correspondence, which is the rigorous criterion used in the Cantorian theory, to compare the cardinality of infinite quantities.

(b) The students tended to think that all infinite sets have the same number of elements. This belief stems from their intuitive understanding of infinity as identical to inexhaustibility.

(c) Many children and adolescents incorrectly assumed that all methods suitable for comparing finite sets are adequate for infinite sets as well. Thus, most of the students who argued that two infinite sets were not equivalent based their claims on the assumption that the maxim "the whole is greater than each of its parts," which is adequate for comparing finite sets, holds for infinite sets as well. This maxim was well-rooted in the students' minds and they expressed a high degree of confidence in it.

(d) The intuitive criteria that the students used to compare infinite quantities were inconsistent with each other and led to conflicting responses and to contradictions of which most of them were unaware. In fact, all but 16% of the students treated each problem separately, and were greatly influenced by the figural context of the problem itself. They justified some of their answers by arguing that all infinite sets had the same number of elements, while justifying other answers with the claim that one infinite set had fewer elements than the other. Only about 8% of them mentioned that they realized the inconsistencies in their responses. One of the students wrote: "All these problems dealt with comparison of infinite sets. When answering some of the problems, I instinctively felt that all infinite sets have the same number of elements, because they all have an infinite number of elements. However, when answering other problems, such as the problem which dealt with a square and a plane, I felt that there are more elements in the plane, since the plane contains the square. But it seems impossible that one infinite set is greater than another infinite set. I realized, after answering these questions, that there are inconsistencies in my answers, and I would like to know if all infinite sets have the same number of elements".

The 16% of the students who were consistent in their answers concerning the comparison of infinite quantities justified their answers to each of the mathematical problems by claiming that all infinite sets have the same number of elements.

(e) There are conflicts between the intuitive criteria that the students used to compare infinite quantities and the formal definitions and theorems of set theory, i.e., between the intuitively accepted statement, "the cardinality of a proper subset is smaller than that of the entire set", and the formal statement, "every infinite set has a proper subset with the same cardinality".

The contradictory and persistent nature of the students' intuitive beliefs in regard to the comparison of infinite sets, as well as the conflicts between these beliefs and the theorems of the Cantorian set theory, is a real challenge for those attempting to teach this theory. In fact, there is evidence, both in the science and mathematics education literature, that contradictory intuitions may be a main obstacle to acquiring formal knowledge (Fischbein & Gazit, 1984; Stavy, Eisen & Yakobi, 1987). Moreover, inadequate intuitive beliefs often continue to affect student's choices of solutions to problems even after formal instruction of the relevant theories (Clement, 1983; McCloskey, 1983). Therefore, instruction of the Cantorian theory of transfinite numbers must take into account the intuitive biases of the learners. It should attempt not only to help learners acquire the definitions and theorems of set theory, but also to assist them in developing efficient secondary intuitions about actual infinity.

3. FIRST STEPS TOWARDS IMPROVING STUDENTS' INTUITIVE UNDERSTANDING OF ACTUAL INFINITY

In what follows we shall first present a learning unit of the Cantorian set theory, which is called: "Finite and Infinite Sets". We shall then describe a study which was aimed at assessing the effects of the unit on students' formal and intuitive understanding of infinity.

3.1 THE "FINITE AND INFINITE SETS" LEARNING UNIT

This unit consists of 20 lessons and is subdivided into four sections:

1. Basic Notions of Set Theory,
2. Equivalence of Infinite Sets,
3. Enumerable Sets,
4. Non-enumerable, Linear Sets.

A special attempt was made, throughout the unit, to interact with the students' intuitive background in regard to infinity and to change their attitude towards their primary intuitive reactions. The following strategies were used to help the students overcome the inner contradictions in their intuitive understanding of actual infinity.

3.2 RAISING STUDENTS' AWARENESS OF THE INCONSISTENCIES IN THEIR OWN THINKING

In our opinion, in order to forestall the use of intuitive methods that are inadequate for comparing infinite quantities in terms of the Cantorian theory, students should realize that these intuitive criteria lead to contradictory answers. Only after they recognize these contradictions, can we proceed to raising their awareness of the need to use rigorous criteria for comparing infinite sets.

Several methods were employed to raise the students' awareness of the inconsistencies in their own thinking about infinity. The most prominent one is the conflict teaching approach based upon Piaget's notion of cognitive conflict (Piaget, 1975). It is aimed at involving students in discussion of and reflection on the inconsistencies in their thinking. Awareness of inconsistencies is expected to lead to a state of inner disequilibrium which can be used to help students resolve the apparent conflicts in their thinking, create new modified concepts and lead to a new equilibrium.

The following example of an activity which makes use of the cognitive teaching approach comes from the section on "Equivalence of Infinite Sets". In this activity the class is divided into teams of four. Each student is asked to answer the following question:

Two sets are given:

$M = \{4, 8, 12, 16, 20, ...\}, \quad N = \{2, 4, 6, 8, 10, ...\}.$

Is the number of elements in set M equal to the number of elements in set N ?
Explain your answer.

Each team is then instructed to discuss its answers and to come to mutual agreement about the correct response. It is likely that in each team some students may claim that "both sets have the same number of elements because they are both infinite", whereas others may argue that "the set M is smaller because it is a proper subset of the set N". Consequently, the participating students will note that both answers seem reasonable and that they are unable to decide which of them is correct.

Then the class is encouraged to discuss the consequences of relying only on intuitive criteria. It is concluded that the 1-1 correspondence may be used to compare the number of elements in both finite and infinite sets.

3.3 DISCUSSING THE ORIGINS OF STUDENTS' INTUITIONS ABOUT INFINITY

It is widely accepted that students' understanding of the sources of their intuitive beliefs is essential to enable them to develop their ability to monitor and control the effects of primary intuitions on their thinking processes. Therefore, the learning unit includes explanations about the sources of relevant intuitive beliefs. For instance, the unit explains that our mental schemes, built as they are on our real life experiences, are naturally adapted to finite sets. We tend to apply these schemes to infinite sets and to accept intuitively generalizations such as: "A proper subset of an infinite set contains fewer elements than the entire set". The tendency to relate properties of finite sets to infinite sets is one of the main sources of the inadequacy of intuitive beliefs with respect to the Cantorian set theory.

Students are also asked questions such as: "What is infinity?" or "How would you explain the idea of infinity to a friend of yours?" Such questions are aimed at eliciting spontaneous responses reflecting the idea that infinity is identical to inexhaustibility, or, the intuitive interpretation of infinity as pure potentiality. It is then explained that this interpretation of infinity is the source of the intuitive belief that two infinite sets are always equivalent, which is inadequate in respect to the notion of transfinite numbers.

3.4 PROGRESSING FROM FINITE TO INFINITE SETS

It is largely acknowledged that infinity can be viewed as an extrapolation of our finite experiences (Tall, 1981; Rucker, 1982; Dauben, 1983). In particular, infinite sets may be seen as an extrapolation of finite sets. Therefore, throughout the unit an effort is made to refer first to finite sets, with which the students are already familiar, and then to deal with infinite sets, discussing the similarities and differences between them.

For instance, the students are asked to solve several problems that deal with a comparison of the number of elements in finite sets. They are instructed to use various methods, such as counting, 1-1 correspondence and the part-whole principle, in order to solve these problems. This activity evokes a number of questions such as:

- What is the basis of each of these methods?

- Can the same methods be used to compare the number of elements in two given infinite sets?

- Is there one general criterion that would enable a comparison of any two sets, finite or infinite?

- What might have led Cantor to choose 1-1 correspondence as the criterion for comparing infinite quantities?

- What are the similarities and differences between the comparison of finite sets and that of infinite sets?

A discussion of these issues may lead to an examination of the adequacy of each of the intuitive methods used by the students to compare infinite sets with reference to the Cantorian set theory.

3.5 STRESSING THAT IT IS LEGITIMATE TO WONDER ABOUT INFINITY

Some of the comments of mathematicians on the puzzling aspects of infinity are quoted in the unit in order to give the students the feeling that it is legitimate to find these aspects perplexing. For instance, we quoted Hahn's comments about the theorem:

An infinite set is equivalent to at least one of its proper subsets. If we look for examples of enumerable infinite sets we arrive immediately at highly surprising results. The set of all the positive even numbers is an enumerable infinite set and has the same cardinal number as the set of all the natural numbers, though we would be inclined to think that there are fewer even numbers than natural numbers.

Later, he suggests his own intuitive explanation of this discrepancy by drawing an analogy between the intuitive definition of cardinal numbers and the discovery of a new species of animals:

This species must be different in some way from the known ones, otherwise it would not be a new species.   (Hahn, 1956, p. 1604)

Comments of other mathematicians, such as Hilbert (1964), Russell (1956), Frankel (1953) and Cantor himself, which are included in the unit, illustrate that these mathematicians were aware that a major breakthrough was needed in the concepts of number, comparison and infinity in order to make the transition from finite to infinite numbers.

3.6 EMPHASIZING THE RELATIVITY OF MATHEMATICS

Yet another strategy is to describe several alternative concepts of infinite numbers developed by different mathematicians such as Bolzano (1950), Cantor (1955), Robinson (1966) and Tall (1980c). Emphasis is placed on analyzing the possible reasons that led each of them to devise his own way of perceiving infinity. Several extensions of the concept of natural numbers are introduced. It is hoped that this exposure will help the students gain a clearer realization of the relativity of mathematics.

3.7 STRENGTHENING STUDENTS' CONFIDENCE
IN THE NEW DEFINITIONS

Students are provided with opportunities to perform mental activities developed with an eye to strengthening their confidence in the new definitions and theorems they have just learned. For example, when dealing with the equivalence between the set of the natural numbers and the set of positive even numbers, students are asked to write down 10 pairs of elements with each containing one element from the set of natural numbers and one from the set of positive even numbers. In this way an attempt is made to offset their tendency to treat each of the sets separately. Rather, they are directed to look at both sets simultaneously by referring to the corresponding pairs. The next step is to guide them to discovering for themselves the formula, $f(x) = 2x$, which describes a 1-1 correspondence between these sets. In the discussion that follows, emphasis is put on the need to use formal methods to determine whether two infinite sets are equivalent, rather than relying on intuitions alone.

4. CHANGES IN STUDENTS' UNDERSTANDING OF ACTUAL INFINITY

In order to examine the impact of the "Finite and Infinite Sets" learning unit on high-school students' understanding of actual infinity, the unit was taught to students aged 15-16 in four tenth-grade classes. Although our main aim was to assess its effects on the students' intuitive understanding of actual infinity, such an evaluation would be meaningless without an assessment of the extent to which the students also acquired the definitions and the theorems which they were taught. Thus, two questionnaires were employed to examine the effects of the instruction on both the students' formal and their intuitive understanding of infinity.

Questionnaire A was designed to check the extent to which students were able to use the concepts and procedures they were taught. It contained 10 mathematical problems which dealt with the definition of equivalent sets, equivalent and non-equivalent sets, and the equivalence between a set and one of its proper subsets. It was administered to the students twice, the first time immediately after instruction and again two months later.

Questionnaire B was designed to assess the effects of instruction on students' intuitions of infinity. It too was administered to the students twice, the first time before instruction and again two months after instruction. This questionnaire contained 14 mathematical problems. In each, two infinite sets were given. The students were asked to determine whether the two sets were equivalent and to justify their claims. At least one of the sets in each problem was two dimensional (i.e., the set of points in a square or the set of points in a plane). Two dimensional sets were not introduced in class during instruction and thus all these problems presented situations that were not dealt with in the learning unit. A brief description of these 14 problems appears in Tables 2 and 3.

The data obtained from Questionnaire A show that 86% of the students acquired the basic concepts, definitions, theorems and strategies necessary to establish the cardinality of and equivalence between infinite sets. These students gave correct solutions to problems that dealt with equivalent and non-equivalent sets, used only formal strategies to justify their claims, and agreed with theorems that contradict maxims which they had regarded as self-evident prior to instruction. The substantial gains made by the students during the instruction phase were maintained over the two-month period between the administration of the two post-tests.

Only 14% of the students used primary intuitive arguments to justify their responses to at least one of the problems in Questionnaire A. Most of these students claimed that an infinite set can not be equivalent to any of its proper subsets and used the part-whole principle to justify this inadequate claim. The intuitive claim that all infinite sets have the same number of elements was rarely used. It is noteworthy that some students reported that they felt uncomfortable with their answers although they knew they were correct. For instance, 8% mentioned that: "It is odd that the set of natural numbers is equivalent to the set of square numbers, which is its proper subset". Similarly, 14% claimed that: "Although the line segment [0,2] is longer than the line segment [0,1], these two sets are equivalent".

Several comments about the extent to which students acquired the concepts they were taught seem appropriate here.

1. After instruction, about 10% of the students incorrectly argued that "a finite set might be equivalent to one of its proper subsets". These students apparently over-generalized the theorem "an infinite set is equivalent to at least one of its proper subsets". A possible source of this misapprehension is our natural tendency to generalize theorems in order to use them in a variety of situations. This difficulty could probably be overcome by a greater emphasis of the differences between finite and infinite sets with respect to the part-whole principle. The idea that properties that hold true for infinite sets may not necessary hold true for finite sets should be discussed.

2. Students' performance on problems involving equivalent sets was better than their performance with non-equivalent sets; the percentage of adequate responses (correct claims and full justification) to problems dealing with equivalent sets ranged from 88% to 97%, whereas the percentage of correct responses to problems dealing with non-equivalent sets ranged from 74% to 80%. One possible explanation for this difference is the fact that the students spent more time and had more practice with enumerable sets. Another possible explanation is that claims of equivalence are verified by direct methods of proof while claims of non-equivalence are verified by the indirect method of proof, which is less familiar to and more problematic for high school students (Roberti, 1987). It is reasonable to assume that explanations about the indirect method of proof, as well as more practice with non-enumerable sets, could improve students' performance on these problems.

In general, our findings indicate that the learning unit on infinite sets may be introduced without particular difficulty starting from the tenth grade. Although some problems were identified, we assume that an improved version of the learning unit would be able to overcome them.

The situation in regard to the effects of instruction on students' intuitions of infinity is far more complex. Tables VIII and IX present students' responses to Questionnaire B.

The Sets Compared A vs B		Before Instruction Cardinality			After Instruction Cardinality	
		Same*	Different	No response	Same*	Different
1. All points in a square	All points in a line	71	24	5	90	10
2. All points in a triangle	All points in an arc	63	37	0	84	16
3. All points in a plane	All points in a line	63	34	3	76	24
4. All points in a square	All points in a line segment	55	42	3	84	16
5. All points in a plane	All points in a circle	52	47	1	76	24
6. All points in a square	All points in a triangle	49	44	7	92	8
7. All points in a square	All points in a larger square	47	50	3	95	5
8. All triangles in a plane	All equilateral triangles in a plane	44	56	0	80	20

* Correct answers

Table VIII : Solutions to Problems Dealing with Equivalent Sets (%)

The Sets Compared A vs B		Before instruction Cardinality					After instruction Cardinality		
		|A|>|B|*	|B|>|A|	diff-erent	Same	No reply	|A|>|B|*	Same	No reply
9. All points in a plane	The natural numbers	48	0	0	48	4	82	18	0
10. All points in a plane	The rational numbers	23	3	0	68	6	83	12	5
11. All points in a line	The "rational" points in a plane	20	0	4	65	11	78	14	8
12. All point in the interior of a circle	The "rational" points in a plane	15	7	2	64	12	84	16	0
13. All points in a line segment	The "rational" points in a plane	6	35	2	50	7	63	29	8
14. The "irrational" points in a plane	The "rational" points in a plane	4	9	8	68	11	90	10	0

* Correct answers

Table IX : Solutions to Problems Dealing with Non-equivalent Sets (%)

The above data indicate that the percentage of adequate responses increased after instruction. The lowest percentage of adequate responses after instruction, were yielded by problems that dealt with the comparison of a linear set and a two dimensional unbounded set (problems 3, 5, 11 and 13). All students who gave inadequate answers to these problems argued that the cardinality of a two dimensional unbounded sets (i.e., the set of all points in a plane or the set of all rational points in a plane) was higher than c, the cardinal number of the set of reals.

The changes in students' responses to problems dealing with a comparison of infinite quantities are also reflected in the criteria they used to determine their solutions. After instruction, the majority of the students made an attempt to implement the formal strategies and theorems that they had learned. For example, on problem 9, 82% of the students claimed that: "We have learned that the cardinality of natural numbers is $\aleph_0$. The cardinality of the set of points in a plane is at least c, since it is at least the same as the cardinality of the set of points in a line. Thus the cardinality of the set of points in a plane is greater than that of the set of natural numbers".

Another strategy used by the students to justify their solutions was that of indicating an analogy between the given problem and one they had discussed in class. For example, in the case of problem 7, 19% of the students claimed that "the problem of comparing the number of points in a square with the number of points in a larger square is similar to that of comparing the number of points in a line segment with the number of points in a larger line segment. The set of points in a line segment is equivalent to the set of points in a larger line segment; thus, the set of points in a square is equivalent to the set of points in a larger square." Only 13% of the students used their former intuitive arguments to justify their solutions to at one or more of the problems in Questionnaire B.

The students' responses to the various problems showed that as a result of instruction the vast majority of them realized that the intuitive criteria they had used when comparing infinite quantities were inadequate in respect to the Cantorian set theory. The students became aware of the need for formal mathematical proof as opposed to intuitive evaluation. Some of their responses clearly illustrate that they became critical about their natural intuitive attitudes. For instance, in solving problem 7, one of the students wrote: "According to my natural reasoning the two sets are equivalent. However, I realize that sometimes my reasoning is misleading. Therefore, I cannot depend only on my natural reasoning and I will try to prove, formally, that the two sets are equivalent." Another student commented, with reference to the same problem, that: "This case is similar to that of the line segment [0,1] and the line segment [0,2]. Therefore, it seems that these sets are equivalent. However, I am not sure since unexpected things happen with infinite sets."

Can one conclude, based on these findings, that as a result of instruction the students developed modified intuitions towards the comparison of infinite sets? Our data do not allow us to draw such a conclusion with confidence. They do not provide evidence to support the claim that ideas that were originally regarded as preposterous by the students, such as the possibility of equivalence between an infinite set and a proper subset of it, now became self-explanatory concept for them. The changes in students' reactions to non-standard situations may indeed be due to modified intuitions, but it is also possible that these changes are a result of the newly acquired formal knowledge of the Cantorian set theory together with an increased awareness of the need to control intuitively based reactions. The fact that some students reported that they felt uncomfortable with their answers although they knew they were correct may be viewed as supporting the claim that as a result of instruction, the students realized that when comparing infinite sets they had to consciously monitor their reactions and not base them on their intuitions, but that their intuitions per se were not modified.

Other data which may be viewed as providing support for this claim derive from the students' responses to mathematical problems involving two dimensional, unbounded sets. As mentioned above, a substantial number of them claimed that a two dimensional, unbounded set contains more elements than a linear set. This may indicate that when dealing with mathematical problems with which they were unfamiliar, the students were still, implicitly, influenced by their primary intuitions. However, it is equally reasonable to argue that students were reluctant to accept the possibility of equivalence between a bounded linear set and a two dimensional, unbounded set because during instruction they had no experience of equivalent sets that differ in dimension. If so, the learning unit may have effectively changed the students' intuitions concerning the equivalence of bounded and unbounded sets, but the change in this specific intuition may not necessarily lead to a change in their intuitions in respect to the equivalence of sets that differ in dimension. It is certainly feasible that eradicating one inadequate intuition does not necessarily eliminate another.

Thus, it seems safe to conclude that the instruction influenced the explicit decisions of the vast majority of the students concerning the comparison of infinite sets. They learned to give conceptually controlled answers rather than spontaneous intuitive ones. A substantial portion of the students developed an "alarm technique" for problems dealing with a comparison of infinite sets. This technique was an upshot of their realization that it is risky to support a mathematical statement by intuitive evaluations alone. These students mentioned that the intuitive criteria they had previously used to compare infinite quantities had led them to contradictions and inadequate solutions. It became apparent to them that their solutions to problems dealing with infinite sets should be based on the formal theorems and strategies that they were taught.

This study shows that the comparison of infinite sets, which illustrates some of the more perplexing aspects of mathematics, can be used to enable students

(a) to accept conclusions which at first appear paradoxical,

(b) to recognize the coercive nature of intuitive thinking,

(c) to understand the need to control their primary intuitions,

(d) to refrain from responding intuitively,

and

(e) to use explicit theorems in order to determine their solutions to the problems.

It is hoped that such instruction would raise students' awareness of the role of intuition in their thinking and affect their attitude towards intuitive responses not only in respect to infinity but also with reference to other mathematical and scientific processes.

5. FINAL COMMENTS

1. Developing a learning unit that takes into account the intuitive background of the learners requires a profound knowledge of the nature of students' intuitions towards the specific mathematical theory. The identification of relevant, inadequate intuitions held by the students is extremely important. This study has shown that in the case of comparing infinite sets, many of the students' primary intuitions were similar to those experienced by mathematicians in the history of the development of the concept. Such palpable parallelism between phylogeny (historical development of the species) and ontogeny (development of the individual), reveals the former as a potential source for identifying students' intuitions.

2. We have already seen that some of the students justified their responses to certain problems by pointing out the analogy between the given problem and one discussed in class. Analogies were not discussed during instruction, yet students used them spontaneously and correctly. Recent studies have shown that analogical reasoning can be useful for helping students achieve conceptual changes (Strauss & Perlmutter, 1986; Clement, 1987; Stavy, in press). The potential of analogies to help students recognize the dissonance in their thinking about infinity and the consequences of including them in instruction examples which illustrate possible correct and incorrect uses of analogies should be explored.

3. In a paper on "Mathematics and the Metaphysicians", Bertrand Russell argues that:

On the subject of infinity it is impossible to avoid conclusions which at first sight appear paradoxical, and this is the reason why so many philosophers have supposed that there were inherent contradictions in the infinite. But a little practice enables one to grasp the true principles of Cantor's doctrine, and to acquire new and better instincts as to the true and the false. The oddities then become no odder than the people at the antipodes, who used to be thought impossible because they would find it so inconvenient to stand on their heads.   (Russell, 1956, p. 1578).

Russell here distinguished between gaining a conceptual knowledge of infinity and acquiring new intuitions. He also claimed that a little practice would enable the learners to acquire both. Our study has shown that the process of acquiring new instincts is not quite that simple. Further, our data indicate that the road from a conceptual grasp of the principles of set theory to acquisition of better intuitions in respect to the comparison of infinite sets may be a long one.

It seems that the various strategies that were used in the learning unit "Finite and Infinite Sets" did indeed enable the students to progress towards acquiring intuitions which are consistent with the theory they learned. However, we lack the means to evaluate these effects systematically. In order to proceed in devising instructional strategies that take into account the intuitive background of the learners we need to develop means to measure "degrees of intuitiveness". A preliminary attempt to measure the intuitive acceptance of a mathematical statement is described in Fischbein, Tirosh & Melamed (1981). Yet, if we believe that the intuitive attitudes of the learner have a crucial effect on his or her concepts and capacity to understand mathematical theories, and if we further believe that intuitions can be modified, we need to devote much greater efforts to devising means that will enable us to base our assessments of the effectiveness of various methods of instruction in modifying students' intuitions on systematic evaluation.

CHAPTER 13

RESEARCH ON MATHEMATICAL PROOF

DANIEL ALIBERT & MICHAEL THOMAS

1. INTRODUCTION

The formulation of conjectures and the development of proofs are two fundamental aspects of a professional mathematician's work. They have a dual character. Firstly there is the personal, intimate side, which aims at clarifying the position the researcher has reached in his/her own understanding, through the statement of explicit hypotheses. Secondly there is the collective side, where a conjecture is proposed for the reflection of other mathematicians, sharing ideas, as yet unsure. In this context a proof is a means of convincing oneself whilst trying to convince others.

These two facets of advanced mathematical thinking are generally absent in undergraduate mathematics at university, where the subject matter is presented as a finished theory, where "all is calm... and certain" and proofs are developed along traditional 'linear', deductive lines.

The epistemology (the understanding of the structure of knowledge) generated by such teaching practices is thus diametrically opposed to the reality of the mathematical community.

A study of textbooks for students at this level appears to confirm that the semantic characteristics of the mathematics – the control of meaning – is not a primary aim. Instead emphasis is placed on the syntactic aspects in carrying out and using the results of algorithms.

This apparent conflict between the practice of mathematicians on the one hand, and their teaching methods on the other, creates problems for students. They exhibit a lack of concern for meaning, a lack of appreciation of proof as a functional tool and an inadequate epistemology.

It may well be that the students' view of whether proof is a necessary mathematical activity, their understanding of the need for rigour, and their preference for one type of proof over another, are concerns which have been neglected by some mathematics educators in favour of a perceived need to preserve the precision and the beauty of mathematics. A consideration of the students' view may be especially important during the transition phase when they are first exposed to the rigour of formal proof as it often occurs in a first year university mathematics course.

Researchers in this area of mathematics education have demonstrated that there is an important difference between communicating sufficient understanding of a proof to convince students of a result, and a formal, rigorous proof that it is true. Balacheff (1982, 1988) has described various levels on which proof may exist, and the importance of distinguishing between convincing arguments and rigorous proof. The latter may well be a suitable instrument to be used in the kind of formal text that mathematicians write in books or research articles, but may not be suitable when initially passing on acquired knowledge to students. One difficulty associated with achieving a proof which is both meaningful and formally acceptable to students is:

> How do we include the main ideas through which we understand *why* the result is true at the same time as the necessary details to make it rigorous?

We shall discuss this and other aspects relating to how understanding of proofs may be better communicated to students. We shall also pay particular attention to studies emphasizing the nature of proof as an activity with a social character, a way of communicating the truth of a mathematical statement to other people, helping them to understand why it is true.

A major area of difficulty linked with this social character of proof which we shall consider here is:

> How can we manage to make students see proof as a necessary step in the scientific process, alongside activities such as research, the formulation of conjectures etc. and not just as a formal necessity required by the teacher, or as an answer given by the teacher in response to a question which the student may not have asked?

These two problems, respectively, have been the subject of research by Leron in the Department of Science Education of Haifa University, Israel and Alibert, Grenier, Legrand and Richard in the Research Group in the Didactics of Mathematics at the University of Grenoble, France. Leron (1983a, 1985a) has proposed a method of structuring proofs to improve the way students understand them, while Alibert *et al* (1986, 1987, 1988abc, 1991), Grenier *et al* (1984, 1985) and Legrand & Richard (1984) have designed a new teaching method involving scientific debates in order to encourage students to see the necessity for proof as a mathematical activity.

2. STUDENTS' UNDERSTANDING OF PROOFS

First we shall turn our attention to the students' perspective of proof in a mathematics course. What are the characteristics of the proofs which they prefer and claim to understand better, and how good is their understanding?

Several researchers, including Fischbein (1982), Movshovitz-Hadar (1988) and Tall (1979) have investigated aspects of the teaching of proofs which may be appropriate for presenting material in a potentially meaningful manner for the learner. There are proofs, for example where the inner parts of the proof are not trivial, where structured proofs and linear (or formal deductive) proofs display similar pedagogical problems.

Tall (1979) is concerned with the students' first acquaintance with proof at university and investigates which of several types of proof they find more understandable. Following Steiner (1976), he suggests the concept of a *generic proof* as a potential way round such problems. Such a proof works at the example level but is generic in that the examples chosen are typical of the whole class of examples and hence the proof is generalizable. This may be contrasted with general proof which works at a more general level but consequently requires a higher level of abstraction. Whilst there may be no replacement for the formal proof from the purely logical point of view, the generic proof may sometimes be preferable if it results in improved understanding on the part of the students.

Discussing the proof of the irrationality of $\sqrt{2}$, Tall describes a study in which 33 first year university students were presented with three proofs of the result: one general, one generic and the 'standard' proof by contradiction. In a second questionnaire, 37 students responded to a generic and a contradiction proof for the irrationality of $\sqrt{5/8}$.

The generic proof as used here was :

We will show that if we start with any rational p/q and square it, then the result p^2/q^2 cannot be 5/8.

On squaring any integer n, the number of times that any prime factor appears in the factorization of n is *doubled* in the prime factorization of n^2, so each prime factor occurs an even number of times in n^2. (For instance, if $n = 12 = 2^2 \times 3$, then $12^2 = 2^4 \times 3^2$.)

In the fraction p^2/q^2, factorize p^2 and q^2 into primes and cancel common factors where possible. Each factor will either cancel exactly or we are left with an even number of appearances of that factor in the numerator or denominator of the fraction. The fraction p^2/q^2 can never be simplified to 5/8 for the latter is $5/2^3$, which has an odd number of 5's in the numerator (and an odd number of 2's in the denominator).

The results showed that the generic proof for the irrationality of $\sqrt{5/8}$ was significantly preferred to the proof by contradiction, both in terms of understanding and lack of confusion. Furthermore there was a highly significant preference for both the generic proof and the proof by contradiction over the general proof of the irrationality of $\sqrt{2}$.

Dreyfus & Eisenberg (1986) gave five proofs of the irrationality of $\sqrt{2}$ to mathematicians who were asked to rank them according to elegance. It is of interest that the experts' personal preferences were for the proofs which were older and more elementary, including a proof along the lines of that described above for $\sqrt{5/8}$. Dreyfus and Eisenberg conclude that clarity and *simplicity* of argument are two principal factors which should guide one when trying to nurture mathematical appreciation. The use of generic examples in proofs may be a way to promote such arguments.

Movshovitz-Hadar (1988) also recommends a "generic-example assisted" type of proof. Applying this method of proof to the theorem:

For any $n \times n$ matrix, n a positive integer, such that the rows form arithmetic progressions with the same common difference d, then the sum of any n elements, no two of which are in the same row or column, is invariant.

She uses an 8×8 matrix as an example:

small enough to serve as a concrete example, yet large enough to be considered a non-specific representative of the general case. The proof for the 8×8 case ... is kind of "transparent", one can see the general proof through it because nothing specific to the 8×8 case enters the proof.

(Movshovitz-Hadar, 1988 p. 19)

It would seem that the explanatory power of such a proof may supplant the generality of general proofs for the student, resulting in more meaningful understanding. Mathematical insight in proof may be more important than precision in these circumstances.

Research by Vinner (1988) shows that when taking this generic approach to proofs, one should be aware that students may be resistant to accepting the proof due to cognitive obstacles arising from their pre-disposition to linear formalism. He gave students two proofs of the mean value theorem which states:

If a function f is differentiable between a and b, and continuous at a and b, then there is a point ξ between a and b such that $f(\xi) = \dfrac{f(b)-f(a)}{b-a}$.

The first proof was the standard algebraic proof, applying Rolle's Theorem to

$$f(x) - \frac{f(b)-f(a)}{b-a}(x-a).$$

The second was a visual proof involving moving the chord AB as shown below, parallel to itself until it becomes a tangent.

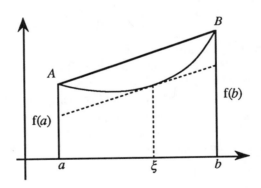

Figure 31 : The mean value theorem

Of 74 students, 29 found the visual proof more convincing, 28 the algebraic proof and 17 considered them of equal value. Those preferring the algebraic method tended to remark that there was something wrong or 'illegal' in the visual approach, and Vinner considered that students develop an algebraic bias through environmental effects to do with 'habit' and 'convenience' rather than cognitive necessity. This feeling of needing a formal deductive proof may emanate from a lack of confidence in any other approach rather than an affinity for the aesthetics of algebra.

Research by Fischbein has uncovered another aspect to the understanding of proof by students. He has found that they may understand the theorem statement itself, they may even, through the use of a structured proof, or otherwise, grasp the structure of the proof, and yet still they may fail to appreciate the universal validity of the statement as guaranteed and imposed by the validity of the proof. This conclusion was reached following a research project in which about four hundred high school pupils with advanced training in mathematics were presented with a correct proof of the theorem:

$n^3 - n$ is divisible by 6 for every integer n.

The students were then given various questions about the validity of the theorem. Whilst 81% checked the proof and claimed it to be correct in every detail, 68.5% agreed with the theorem and 60% considered the generality of the theorem guaranteed by the proof, only 41% of the students accepted all three of these. Further only 24.5% accepted the correctness of the proof and at the same time answered that additional checks are not necessary, and only 14.5% were completely consistent in their answers.

To a mathematician the proof of a theorem is

the absolute guarantee of the universal validity of the theorem. He believes in that validity.
(Fischbein, 1988, p. 17)

The question is how does one convey, in a proof or otherwise, the information necessary for the individual student to synthesize cognitively the formal understanding of the truth of the result and an acceptability of its universal validity? This pedagogical necessity in mathematical proof exposition may be one which still needs to be addressed.

3. THE STRUCTURAL METHOD OF PROOF EXPOSITION

Mathematical proofs have long tended to have a format which requires them to be read in a strictly serial/sequential manner, with sub-proofs, or lemmas, which are themselves also strongly sequential. Such a style of proof makes the acquiring of a global over-view something which requires sufficient mathematical sophistication to understand the details of the sequence well enough to be able to relate them to the overall theme as one progresses through the proof. Such an ability to switch as and when necessary from a sequential view of the mathematics to a global one and vice-versa is a characteristic of one who has been described as a *versatile learner* (Tall & Thomas, 1989). In order to promote versatility in students, Tall & Thomas (1990) have highlighted the importance and value of actively encouraging a global view of the mathematics, indeed promoting it in one's teaching in addition to the more familiar serialist presentations of mathematics. Using the benefits provided by the computer paradigm they have obtained some evidence that a versatile learner, who is able to switch between a global and a sequential view of the mathematics, is more likely to be successful in the early learning of algebra as generalized arithmetic (with 12–14 year old pupils) and in the initial stages of the calculus, and have placed this improved ability in a theoretical context applicable to other areas of mathematics.

Leron (1983a, 1985a) has attempted to fuse formal and informal methods of presentation into a proof which is rigorous and yet explanatory. The two informal practices (heuristics) he has built into the formalism are:

- the prefacing of a long, complex proof with a short, intuitive overview,
- a method of constructing a mathematical object, a solution-object, to satisfy a system of constraints by using the given constraints to search for the solution-object and then using its form to define it.

This kind of proof he calls a *structural proof*.

Here the primary aim is not merely to convince, but to help the listener or reader gain a real understanding of the ideas behind the proof and its connections with other mathematical results. In comparison, the usual 'linear code' type of proof not only often fails to elucidate the main ideas, but may even obscure them. This type of proof may well be suited to ensuring the validity of a proof, but it is unsuitable for the role of mathematical communication. It has even happened, in some extreme cases, that the author recognizes that his/her own proof fails to give any real insight into the understanding of the mathematics. For instance Deligne, one of the most famous contemporary mathematicians (and winner of a Fields Medal), wrote after a very formal proof about derived functors and categories,

"I would be grateful if anyone who has understood this demonstration would explain it to me" .
(Deligne, page 584)

Clearly proving and explaining seem to be two different kinds of mathematical activities.

The linear formalism of traditional proof may be described as the minimal code necessary for the transmitting of the mathematical knowledge. It appears, however, that in several important respects, it is a sub-minimal code, resulting in an irretrievable loss of information vital for understanding.

Whilst most of the work in mathematics education rightly seeks to improve the learning and communication of mathematics by supplementing the formalism, it is also important to look at the formalism itself and consider how it too might be improved, leading to better communication and understanding. It is certainly to be hoped that students of mathematics are actively engaged in discovering and constructing as much of the mathematics they learn as possible, but it is also necessary to find better ways of communicating the products of such mathematical activity to others.

The fundamental concept underlying the structural method of presenting proof is to arrange the proof in levels, proceeding from the top down. Each level consists of short autonomous modules, each embodying one main idea of the proof. This type of structure is already recognized and well-known in computer science as a method of structuring complex computer programs, where it is called top-down programming.

It should be noted that the term 'main idea' which we here refer to is used more in the abstract sense of that idea which enables one to gain an overview of a sub-section of the proof rather than indicating that which is mathematically among the most important ideas in the proof.

It is useful to analyze some examples of this type of structural proof (see below) in order to ascertain the difference in the treatment given by the two methods and to see specifically what is meant by the 'main ideas' of a proof. A main idea is often contained by the construction of a new, intermediate object called the *pivot* (so named because the rest of the proof hinges upon it), designed to mediate between the hypotheses and the conclusion. The pivot occupies a central position in the proof (or sub-proof), and so it offers a vantage point from which one may view the global architecture of the proof. In the linear approach, the pivot is often poorly treated and its potential for improving understanding wasted. Rather the proof begins to resemble the pulling of a rabbit from the hat, since the pivotal concept may be introduced near the beginning of the proof, possibly by simply a bare statement of its definition.

In the first level of the proof one *tool-pivot* is identified (e.g. set, relation, function etc...), the existence of which is essential for the proof to be developed. Since this is to be a tool, it is then given some properties which are also to be used in the proof, although the actual existence of the tool-pivot is not proved at this stage of the proof, but at a later, deeper one.

In the second level of the proof the tool-pivot becomes an *object-pivot* which is to be constructed, subject to certain constraints. An heuristic discussion follows concerning the possibility of achieving the construction under the given constraints. This construction itself may, if necessary, be further divided up and treated on several levels. Proceeding in this way avoids the view that the pivot is just a construction which in the eyes of the student is 'an extremely clever answer to a question which was never asked'. These concepts are probably best understood by considering some examples of structural proofs and their 'linear' equivalents. We shall consider here two of the examples given by Leron (1983a).

Theorem : There exist infinitely many triadic primes
(i.e. numbers of the form 4k + 3, for integral k)

3.1 A PROOF IN LINEAR STYLE

Consider the product of two monadic numbers :

$$(4k + 1)(4m + 1) = 4k.4m + 4k + 4m + 1 = 4(4km + k + m) + 1$$

which is again monadic. Similarly, the product of any number of monadic primes is monadic.

Now assume the theorem is false, so there are only finitely many triadic primes, say $p_1, p_2, \ldots, p_n$. Define

$$M = 4p_1 p_2 p_3 \ldots p_n - 1.$$

If $p_i \mid M$ then $p_i \mid 1$ since $p_i \mid 4p_1 p_2 p_3 \ldots p_n$. Since this is impossible, we conclude that no p_i divides M. Also 2 does not divide M as M is odd. Thus all M's prime factors are monadic, hence M itself must be monadic. But

$$M = 4p_1 p_2 p_3 \ldots p_n - 1 = 4(p_1 p_2 p_3 \ldots p_n - 1) + 3$$

is clearly triadic – a contradiction. Thus the theorem is proved.

We note that in the above proof the general plan is never revealed, neither is the purpose of the various steps taken. In the absence of any explanation as to why certain steps are taken (such as considering the product of two monadic numbers) the student may be reduced to merely checking the validity of the deduction at each step.

3.2 A PROOF IN STRUCTURAL STYLE

Level 1 – Suppose the theorem is false and let p_1, p_2, p_3, ..., p_n be all the triadic primes. We construct (in level 2) a number M [the pivot] having the following two properties :

   (a) M as well as all its factors is different from p_1, p_2, p_3, ..., p_n,

   (b) M has a triadic prime factor.

These two properties clearly produce a contradiction, as we get a triadic prime which is not one of p_1, p_2, p_3, ..., p_n. Thus the theorem is proved.

In the Elevator – (a metaphor for the process of descending in levels)

How shall we approach the definition of M? In the light of Euclid's classical proof, it is natural to try $M = 4p_1 p_2 p_3 \ldots p_n + 1$. This indeed meets requirement (a) but not (b). In fact, since for all we know M itself may turn out to be prime, it must be triadic to meet (b).

Thus a natural second guess is $M = 4p_1 p_2 p_3 \ldots p_n + 3$. However, this has another "bug" – since one of the p_i's is 3, M is divisible by 3, in violation of (a). But this bug, once discovered, is easy to fix – simply eliminate 3 from the product in the bugged definition.

Level 2 – Let $M = 4p_2 p_3 \ldots p_n + 3$ (we assume $p_1 = 3$). We show that M satisfies the two requirements from Level 1.

Requirement (a) means that no p should divide M. Indeed, p_2, p_3, ..., p_n do not divide M as they leave a remainder of 3; and 3 does not divide M as it does not divide $4p_2 p_3 \ldots p_n$.

As for requirement (b), suppose on the contrary that all of M's prime factors were monadic. Then M, as a product of monadic numbers, would itself be monadic (Lemma, Level 3) – a contradiction. Thus (a) and (b) are satisfied.

Level 3 – Lemma. A product of monadic numbers is again a monadic number. (The proof is given above.)

We note here that the top level 1 gives a global view of the proof. The 'elevator' affords the opportunity for informal discussion within the proof, including part of the process of finding a proof. The necessary lemma is only introduced when it is needed, avoiding its introduction at the start of the proof with no mention as to its later use.

As a second example, Leron considers the proof of the theorem:

$$\text{If } \lim_{x \to a} f(x) = L \text{ and } \lim_{x \to a} g(x) = N, \text{ then } \lim_{x \to a} f(x)g(x) = LM.$$

A standard text-book proof starts with an $\varepsilon > 0$ and shows by a number of intermediate steps how a $\delta > 0$ can be found so that, when $|x - a| < \delta$ we have $|f(x)g(x) - LM| < \varepsilon$. At level 1, Leron starts with $\varepsilon > 0$ and assumes that the pivot, $\delta > 0$, can be found at level 2, thus proving the theorem subject to the level 2 construction. He then sets out to find such a δ, by looking at what he is trying to achieve, using the inequality

$$|f(x)g(x) - LM| = |L(g(x) - M) + M(f(x) - L) + (f(x) - L)(g(x) - M)|$$

$$\leq |L|.|g(x) - M| + |M|.|f(x) - L| + |f(x) - L|.|g(x) - M|$$

to break this down to level 3 problem seeking to bound the terms in the second line.

After a process of trial and improvement to achieve this, he is able to take the 'elevator' back through the levels to complete the proof.

The format for structural proofs suggested by Leron and seen in these examples is:

1. Introduce the pivot as a system of constraints, i.e. define it implicitly by postulating its properties.

2. Without actually solving the system, use the pivot as introduced in step 1 to derive the conclusion of the theorem.

3. Discuss heuristically the solution of the system to find how the pivot might be constructed.

4. (Recursion step). Solve the system, repeating steps 1–4 if necessary. That is construct (or prove the existence of) the pivot then prove that it satisfies the postulated properties. If some of the sub-proofs are themselves complicated, introduce sub-pivots and repeat the four step procedure.

(Leron, 1985a, p. 12)

Though such proofs are clearly longer than standard linear type proofs, the claim for them is that loss of economy is more than balanced by a gain in understanding and learning, and that it would be better to leave out some of the low-level details from these structural proofs, if brevity is required, than to exclude the high-level ideas and connections left out by linear, deductive expositions.

It may be argued that the knowledge of the structure of the proof, as illustrated above, is best left for the students themselves to discover, however experience shows that this kind of task is beyond the capability of most undergraduates with a standard mathematical training. They are simply unable to decode the proof and are reduced to meaningless manipulation of the formal code itself, with no awareness of the ideas and concepts it represents. One of the goals of the structural method is to train students to structure linear proofs. Using the structural method of proof also leads to the possibility of several new structure related activities which encourage the learner to reflect on the process of theorem proving itself. For example they may

- complete the lower levels of a proof, given the higher levels.

- take a standard proof and examine its structure (not an easy exercise but one which may result in deeper understanding of the proof).

- examine the depth of similarity of two theorems which exhibit some similarity in terms of the levels of the theorems.

The major difference between the approach outlined above and the traditional linear proof style is that the students are given a means of understanding the choices that, generally, the teacher presents without any indication that there had actually been a choice involved. Previously some questions about the choices made may have arisen in the minds of some alert students (although no answers would generally be provided during the presentation of the proof) but for many of them understanding a proof is synonymous with merely checking in a sequential manner the validity of the deduction at each step, much as a computer might execute a program. Unable to construct any personal meaning for proofs like these, even simple ones, many of them must feel either cheated or stupid, and certainly they are not in a good position to further develop their scientific capability. In contrast, the understanding gained by the students from a structural type of proof may lead to real scientific autonomy on their part.

4. CONJECTURES AND PROOFS – THE SCIENTIFIC DEBATE IN A MATHEMATICAL COURSE

A further step in the direction of scientific autonomy is taken by the Grenoble school (Alibert *et al*, 1986), attempting to enable students to see proof as a necessary part of the scientific process of advancing knowledge, rather than just a formal exercise to be done for the teacher. An experimental teaching method, set in a theoretical framework, was devised and applied to the teaching of mathematics in the first year of university. The theoretical framework was based on the following general cognitive and didactic hypotheses:

1. Constructivism - the theory of knowledge acquisition in which students construct their own knowledge through interactions, conflicts and re-equilibrations involving mathematical knowledge, other students and problems. The interactions are managed by the teacher who makes the fundamental choices.   (Brousseau, 1986)

2. Knowledge is made firmer when it has been constituted and applied in more than one

appropriate conceptual setting.   (Douady, 1986)

3. The role of contradictions – how they may be made sharper and more explicit and how they may be resolved.   (Balacheff, 1982)

4. The importance of the role of a group of students in the construction of individual meaning (Bishop, 1985; Balacheff & Laborde, 1985)

5. The influence of meta-mathematical factors such as the systems of representations and how these may be worked on explicitly in order to emphasize teaching points.

6. The construction of a learner's epistemology which includes the set of problems, situations, etc, which in the student's mind give meaning to the concept through their association with the introduction and progressive constitution of the concept.

The experiment took place at the University of Grenoble with sections of about one hundred students (95 of the first year and 130 of the second), all of whom were in the first year of a programme called DEUG A (First year university students all taking courses in mathematics, physics and chemistry). The lessons were given to the whole of a section in an ordinary lecture theatre and the experiment lasted the complete year with the group.

4.1 GENERATING SCIENTIFIC DEBATE

The classroom teaching was built around several new customs. Uncertainty in the learning place is important and room should be left for it. In mathematical knowledge this uncertainty is institutionalized in the notion of conjecture, and in this study the validation, and even the production of these, was devolved to the community of students. They were required to produce and validate conjectures relevant to their mathematics curriculum. Underpinning this custom was the principle that the functional nature of proof only arises in situations where students meet the uncertainty of mathematical propositions.

This worked in the classroom as follows:

- *First step*: the teacher initiates and organizes the production of scientific statements by the students. These are written on the blackboard without any immediate evaluation of their validity.

- *Second step*: the statements are put to the students for consideration and discussion. They come to a decision about their validity by taking a vote, with each opinion supported in some way, e.g. by scientific argument, by proof, by refutation, by counter-example, etc.

- *Third Step*: the statements which can be validated by a full demonstration become theorems, whilst those which are established as incorrect are preserved as "false-statements", with a corresponding counter-example.

The demonstrations are produced through interaction between the students and, when necessary, the teacher, after the students have been confronted with the particular problem during a debate. In this form of 'scientific debate' the arguments forming the proof are not addressed by the students to the teacher, but rather to the other students. We have to distinguish here between 'proofs to convince' someone (such as another student) of something that is not already a part of his institutionalized knowledge and 'proofs to show', where the aim is to show someone (such as the teacher) that one has reached some knowledge that he/she already possesses. One of the main hypotheses of research is that the activity involved in the first process is fundamentally different from that involved in the second, and is able to produce a deepening of knowledge and its meaning. The theoretical description of such a teaching system, used since 1984, is based on this hypothesis and is referred to as a 'codidactic situation' (Alibert, 1991). This is a situation in which the student seeks to convince both himself/herself, and others at the same time, of the truth of a conjecture formulated in response to a problem which the whole group is trying to solve. The students are all aware that the conjecture is not necessarily true and in particular that it is not yet an established part of institutionalized knowledge.

4.2 AN EXAMPLE OF A SCIENTIFIC DEBATE

A scientific debate of the type described above starts with a statement such as in the following actual example :

If I is an interval on the reals, a is a fixed element of I, and x an element of I, then we set, for f integrable over I,

$$F(x) = \int_a^x f(t) \; dt.$$

The teacher then asks the question :

"Can you make some conjectures of the form : if f ... then F ...?"

In response to this, in the example given by Alibert (1988a), about 20 statements of varying complexity were produced by the students. One of the sessions started with an examination of the first of these.

"If f is increasing then F is increasing too" (which happens to be false).

The variation of F is one of the items from the curriculum that the students had to learn. During this session the following steps of the proof construction were observed:

(a) *Counter-example* : one student produced an example of a function f contradicting the statement. The class then concluded that the statement was false.

(b) *Statement Modification* : another student proposed that "If f is monotonic then F is monotonic too." (This, of course, is also false.)

(c) *Counter-example* : the same function as used before under (a), but defined on a different interval, contradicted the modified statement too.

(d) *Observation* : by considering the counter-example it seemed that if f ≥ 0 then F is increasing.

(e) *A New Conjecture* : A student now proposed that "If f ≥ 0 then F is increasing." (The majority of the students thought that this statement was false though it is of course true.)

(f) *An Argument* : the student produced the explanation:

$$F(x') - F(x) = \int_x^{x'} f(t) \, dt \geq 0 \text{ if } f \geq 0 \text{ and } x' > x.$$

Interestingly, many of the students did not believe that this was always true and this stage of the debate revealed to the teacher that some of the results and definitions, previously discussed and settled, had been misunderstood by many of the students. In particular they had not fully understood the convention that

$$\int_x^{x'} f(t) \, dt \text{ is the Riemann integral on the segment } [x, x'] \text{ if } x' > x,$$

and minus the integral if $x' < x$.

Analysing this phenomenon we may say that this was a reappearance of old, stable knowledge about the integral learned in previous years.

(g) *Validation* : the students reached a validation of the argument of (f) with:

$$\int_x^{x'} f(t) \, dt \geq (x' - x)\inf(x), \text{ in this case.}$$

This debate took up most of a two-hour class.

The above example illustrates that the propositions debated during the sessions were far from trivial. They also allow students to tackle real problems involving important concepts. Even though many of the statements considered are false they are still very important because, firstly, they discover what students really think about a concept at that precise point in the course and, secondly, the debate about their proposals enables students to be convinced of any false ideas or deep misunderstandings of the concepts which they may hold.

This experience teaches them that proof is really a tool which may be used to improve ideas and separate false intuition, however natural it may appear, from true mathematical statements; to communicate and hence validate or refute mathematical ideas.

4.3 THE ORGANIZATION OF PROOF DEBATES

The organization of the type of debate outlined above involves the use of some precise techniques, such as:

> *Initialization time*– during the first lesson of the year the teacher outlines the way that the course will proceed, partly through explanation and partly through illustrating the process using a debate on a simple mathematics problem. The teacher initiates this by asking a question and a debate about the validity of the answers proceeds, without specific rules.
>
> *Reinforcement time* – There should be two or three of such lessons at the start of the course, with rules for the debates progressively introduced. Some are simple ones – for example speak loudly; speak to one another; listen to the other students – whilst others are more subtle.
>
> *The teacher's position* – The teacher has a precise role to play right from the first lesson. If he/she faces a question which should be considered by all the students then he/she should ask for a conjecture to be produced. The conjectures produced in answer to the problem set should be written on the blackboard without comment. After allowing two or three minutes for reflection the teacher asks for a vote. The students are asked to vote true, false, can't decide or refuses to decide on each conjecture. Then each opinion has to be supported by mathematical arguments.
>
> In this way the students as a group learn that the formulation of conjectures is a useful and necessary activity and that to make mistakes is a normal stage in the learning process.
>
> *The rules* – It has been observed that, at this stage in the course, the teacher is frequently called on to close a debate by expressing his/her opinion on the question in hand because the students have been unable to agree. This inability to convince one another needs addressing and a special lesson called the circuit is introduced. The aim of this lesson is to give the students the means to refute statements. Students are told that: "In mathematics a statement is true if and only if it has no counter-example." This lesson uses some very simple situation to produce conjectures, to refute them and so gradually build up the rules used in mathematics. It is very important here that the mathematical context does not hide the logical problems.

After this session the counter-example should become a very powerful tool for students to use to refute a statement or, more generally, to understand a particular proof. For instance:

> To express in mathematical form that a function F *does not have* an 'infinite positive limit' as the variable becomes infinite is not a simple exercise for students at this level. They do know, however, how to express the idea that F *has* an 'infinite limit' as x becomes infinite (positive), namely :
> For every real A there exists B such that if $x>B$ then $F(x)>A$.

Some have also learned that to negate such a statement one 'replaces *for every* by *there exists* and vice-versa', but this becomes merely a meaningless manipulation of formal code, as we have discussed above. The understanding of this manipulation becomes clear if it is presented as the formulation of the existence of a counter-example for the previous statement.

4.4 EVALUATING THE ROLE OF DEBATE

As a result of this teaching method, concerning the role of proof in mathematics, and the formation of scientific truth, it has been observed that erroneous ideas are no longer considered by the students as faults but as a normal scientific event, and a productive one at that. The students in the study described above were given a questionnaire about the comparison of teaching methods. Many of the students replied that they preferred the course incorporating debates and emphasized that they allowed them to understand the problem which the new mathematical knowledge was aimed at solving, and also what errors may be made too. Some of the comments about the scientific debates from the questionnaires were :

"It compels us to reflect more on the question. One often listens to a clear lecture without reflecting deeply."

"A concept introduced through some conjecture makes the problem that the concept poses much clearer than in a lecture."

"It allows us to have several views, to eliminate some intuitive ideas that are wrong."

Many students also stated how difficult they found the study of a conjecture to be, involving as it does the stating of the problem, the construction of the proof and the formulation of ideas when one is uncertain about the truth of a proposition.

"For me the hardest thing is to find counter-examples when I think that a conjecture is false."

"I often have difficulty forming an opinion, and following it up with a proof."

Certainly the students felt involved and interested and, without excluding the value of 'traditional' lectures, found the debates very useful.

5. CONCLUSION

In this chapter we have looked briefly at some of the research into mathematical proof and its presentation. We have considered different methods of presenting proofs in order to improve students' understanding, including generic proofs and structural proofs. We have also looked at the environment within which these proofs are examined by students with particular regard to the scientific debate as an alternative to traditional presentations.

Amongst the conclusions which some researchers are reaching with regard to students' perceptions of the importance of proof and their understanding of individual proofs are:

(a) There should be a distinct and important difference between the kind of proofs produced by a mathematician researching new areas of mathematics in order to convince others that they have indeed broken new ground and the proofs of these results which will later be used to transmit the results to students of higher mathematics. The latter proofs may need to have extra material included which gives a global view of the proof and its structure if they are to be meaningful to the average student and not just a linear sequence of symbolic reasoning whose step-by-step validity is to be checked.

(b) The context in which students meet proofs in mathematics may greatly influence their perception of the value of proof. By establishing an environment in which students may see and experience first-hand what is necessary for them to convince others, of the truth or falsehood of propositions, proof becomes an instrument of personal value which they will be happier to use in future.

Whatever the cognitive benefits of these approaches to students' intellectual appreciation of specific proofs, and of proof as a mathematical tool, the problem remains of transmitting the methodologies to other teachers so that they may investigate their effectiveness for themselves. The reader might thus like to consider his/her role in this, and review

(i) students' reactions to the proofs presented. Do they display both appreciation for them and understanding of them?

(ii) how proofs which (s)he currently teaches could be restructured to present a potentially more meaningful face to students.

(iii) how a topic/course which (s)he currently teaches could be developed along the lines of a proof debate in the light of the methodology described above, to promote greater appreciation of the necessity of proof in mathematics.

Such a self-appraisal of one's current practice could be very valuable both from the point of view of one's own appreciation of the results and the understanding of one's students. Those readers not so closely involved in these aspects of mathematics education (and others) may like to assess their own understanding in depth of a proof which they are familiar by re-writing the proof in a structural form. Alternatively (and rather more testing) it might be more interesting to attempt to produce structural proofs of one or more of the following results from the usual formats given in standard texts:

1. Cantor-Bernstein Theorem – Let A and B be two sets, If there exists one-to-one maps from A into B and from B into A, then there exists a one-to-one map from A onto B.

2. A connected manifold is path-connected.

3. Sylow's Third Theorem – The number of Sylow p-subgroups of a finite group G is equal to 1 modulo p.

ADVANCED MATHEMATICAL THINKING
AND THE COMPUTER

ED DUBINSKY AND DAVID TALL

1. INTRODUCTION

The computer can be used as a tool to complement advanced mathematical thinking in a variety of ways. In research it has been used to provide data to suggest possible theorems, to seek counter-examples and to carry out onerous computations to prove theorems involving only a finite number of algorithmic cases. In education it can be used for the same objectives, and for one other major purpose: to help students conceptualize, and construct for themselves, mathematics that has already been formulated by others.

There are already many computer tools available for general use. Symbolic manipulators have been used in research, but with less initial success in education. We hypothesize that success using the computer in education is enhanced by using the computer for explicit conceptual purposes and report empirical research which supports this hypothesis. New software environments are being developed which enable the student to explore concepts in a directed and meaningful way, and which suggest new approaches to mathematics more appropriate for the learner.

Programming can be used to support both mathematical research and mathematics teaching. But when it is simply added to the curriculum without very specific aims in mind it has not always been successful. We will discuss the way in which a computer language, designed so that the programming constructs mirror mathematical constructs, can assist students to carry out mathematical processes and encapsulate them as mathematical concepts.

2. THE COMPUTER IN MATHEMATICAL RESEARCH

Mathematical research passes through several distinct stages of development, from the germ of an idea to the formalities of proof:

In Mathematics, as in the Natural Sciences, there are several stages involved in a discovery, and formal proof is only the last. The earliest stage consists in the identification of significant facts, their arrangement into meaningful patterns and the plausible extraction of some law or formula. Next is the process of testing this proposed formula against new experimental facts, and only then does one consider the question of proof. (Atiyah, 1984)

Computers have proved useful in every stage of this development. In the initial exploration phase computer generated data has led to surprising new intuitions and new theory. The famous example is that of Lorenz, studying the outcome of differential equations to predict the weather, who wished to repeat a cycle of events to analyse it in greater detail. Instead of starting from the beginning of a run, he took numbers occurring part way through a previous run and found, to his amazement, that the subsequent pattern diverged enormously from his previous data. He then realized that the output of the previous run had given numbers only to three places: 0.506 instead of the internally stored number 0.506127. The small variation in initial conditions had given a large variation in long term behaviour – knowing initial conditions in a practical sense cannot be used to predict the eventual outcome and *chaos theory* was born (Lorenz, 1963).

Since that time, sensibly programmed environments have proved increasingly valuable to produce data to suggest possible conjectures. Recent developments in the theory of iteration of functions, leading to the beautiful fractal pictures that have become well known even to the general public, arose from research begun, but abandoned, in the earlier part of this century because of the massive computations involved. It was only with the arrival of the computer that the results of the computations could be represented graphically, leading to surprising pictures and new hypotheses to be tested first by drawing, then by a search for formal proof. Likewise, in the theory of dynamical systems, computer graphics have exhibited phenomena that might not have otherwise come to light. Software for the investigation of such phenomena is now generally available. For instance, figure 32 shows a model of a possible orbit of a tiny satellite round two larger bodies, alternately oscillating between revolving round one then moving into a position of superior gravitational pull of the other and moving, for a time, to revolve round the other (Koçak, 1986). It is interesting to note that this book features a significant number of research problems for which there is a clear visual idea of possible solutions but for which no formal proof was available at the time of publication. The theory of dynamical systems and chaos is a paradigmatic example of a new branch of mathematics in which the complementary roles of computer generated experiments to suggest theorems and formal mathematical proof to establish them with logical precision go hand in hand.

Chaos has become not just a theory but also a method, not just a canon of beliefs but also a way of doing science. Chaos has created its own technique of using computers, a technique that does not require the vast speed of Crays and Cybers but instead favours modest terminals that allow flexible interaction. To chaos researchers, mathematics has become an experimental science, with the computer replacing laboratories full of test tubes and microscopes. Graphic images are the key. "It's masochism for a mathematician to do without pictures," one chaos specialist would say. "How can they see the relationship between that motion and this, how can the develop intuition?".
(Gleick, 1987, pp. 38–39)

In the second stage of mathematical thinking, where conjectures have been made more precise and serious attempts are being made to test them, computers may be used sometimes to generate appropriate examples or counter-examples. Nearly two centuries ago, after a prodigious number of calculations, Euler formulated the conjecture that a sum of at least n positive nth powers of integers are required to produce an nth power. So forbidding were the calculations required to investigate this that it stood without proof or refutation until a computer search in 1969 by Lander and Parkin produced the counter-example:

$$27^5 + 84^5 + 110^5 + 133^5 = 144^5.$$

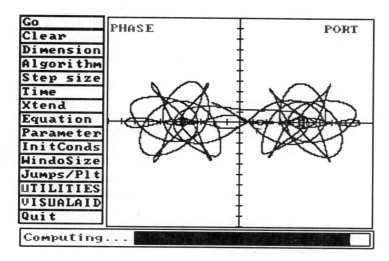

Figure 32 : Chaotic movement of a satellite round two larger bodies

This case was fortunate, in that the discovery of a counter-example showed the conjecture to be false. On the other side of the coin, the inability to find such a counter-example will not show a conjecture to be true. Goldbach's conjecture, that any even number greater than two is a sum of two primes, remains unproven, even though computers have found an appropriate decompositions into two primes for all even numbers up to a formidable size.

In 1916 Bieberbach conjectured that an analytic function

$$z + a_2z^2 + ... + a_nz^n + ...$$

which was 1–1 on the unit disc satisfied

$$|a_n| \leq n.$$

Bieberbach proved the case $n=2$, but by the early 1980s, only the cases up to and including $n=6$ had been proved, by a variety of different methods. Louis de Branges worked for seven years and in 1984 developed a technique which proved the Bieberbach conjecture subject to a condition that could be checked algorithmically. A colleague, Walter Gautschi, ran the method on the Purdue university super-computer – one of only three in the United States at the time – and verified the method as far as the 25th coefficient. The computer proved a vital confirmation at a difficult time for de Branges who had previously twice published erroneous proofs of theorems and found his latest and most complex deductions considered suspect by the mathematical community. His proof was subsequently vindicated when the final steps were confirmed by other means (Kolata, 1984).

In the final stage of mathematical thinking, when a formal proof is being sought, the computer may prove decisive when the question can be reduced to a finite number of cases, each which can be investigated algorithmically. The most famous example is the four colour problem, which Appel & Haken (1976) reduced to a finite (but large) number of alternatives which were resolved by computer. Now the computer is being widely used in combinatorial problems in group theory, algebraic geometry, and other areas with an algorithmic content that can be programmed, leaving the computer to carry out the complex calculations.

The proof of the four colour theorem raises a significant issue in advanced mathematical thinking. For, although there is an apparently impeccable logic in the listing of the possibilities and their checking by computer, the proof itself seems to shed no light as to *why* the theorem is true. Some mathematicians are happy with the situation. For them the process of proof is a mechanistic sequence of deductions from axioms and it is important that, in the actual proof process itself, there are no intuitive leaps that are not subject to logical scrutiny. The logic of the computer is for them an acid test.

However, others involved with mathematical research sense the need not only for the security of logical deduction from a proof, but also some kind of insight as to how the concepts fit with other known results. Without such insight there is always for them the insecurity that some small logical error may be found which renders the argument fallacious. Without some overall view of the pattern there may be a distinct lack of vision as to the possible direction of future research. And, given the ever growing complexity of computer software, there may be errors in the programming which, if the principles are not fully understood, may lead to precisely the weak links that those requiring only a logical approach may fear.

Thus there is value in using the computer to complement the human creative thinking process both in providing environments for exploration into possible new theorems and also to carry out algorithmic calculations to provide mathematical proof, but it is necessary to acknowledge that such methods have weaknesses as well as strengths.

3. THE COMPUTER IN MATHEMATICS EDUCATION – GENERALITIES

All the various ways that computers are used in research are potentially available for teaching and learning advanced mathematics. For example, students may learn to program in order to tackle certain types of problem, or they may use general purpose software as an environment to explore ideas. The main difference between the activities of undergraduate students and mathematical research is that the former usually covers knowledge domains which are known to the more experienced members of the mathematical community, whereas research is attempting to break new ground. Of course, *to the student* the mathematics is new, and here there may be strong analogies with research, but the far greater portion of a student's work is concerned with mathematics that is already part of an organized knowledge system. This opens up a further possibility for the use of the computer in mathematical education, through the development of computer software designed to help the student conceptualize mathematical ideas.

Recent research into concept development shows consistently the complexity of an individual's mental imagery: students can give the "right" answers for the wrong reasons, whilst "wrong" answers may have a rational origin. In particular, many researchers have realized that student errors are often the product of misconceptions brought about using old knowledge in a new context where it no longer holds good. This leads to the hypothesis that learning may be improved by helping students *construct* knowledge in their own minds in a context which is designed to aid, or even stimulate, that construction. One way of doing this is through providing richly endowed computer software which embodies powerful mathematical ideas so that the student can manipulate and reflect on them. Another is to have the student program mathematical constructions in a computer language designed so that the act of programming parallels the construction of the underlying mathematical processes.

A computer can also give much-needed meaning to mathematical concepts that students may feel are "not of the physical world" but in the mind, or in some ideal world. It is generally agreed that ideas are easier to understand when they are made more "concrete" and less "abstract". When an abstract idea is implemented or represented in a computer, then it is concrete in the mind, at least in the sense that it *exists* (electro-magnetically, if not physically). Not only can the computer construct be used to perform processes represented by the abstract idea, but it can itself be manipulated, things can be done to it. This tends to make it more concrete, especially for the person who constructed it. Indeed, it is in general true that whenever a person constructs something on a computer, a corresponding construction is made in the person's mind. It is possible to orchestrate this correspondence by providing programming tasks in an appropriate programming language designed so that the resulting mental constructions are powerful ideas that enhance the student's mathematical knowledge and understanding. Moreover, once the various constructions exist on the computer, it is very useful to reflect on what they are (in terms of how the computer makes them) and what processes they can engage in.

4. SYMBOLIC MANIPULATORS

The use of symbolic manipulators has powerful advocacy from several quarters. Lane *et al* (1986) suggests ways in which symbolic systems can be used to discover mathematical principles and Small *et al* (1986) reports the effect of using a computer algebra system in college mathematics. In the latter case the activities often consist of encouraging students to apply a technique already understood in simple cases to more complicated cases where the symbolic manipulator can cope with the difficult symbolic manipulations.

However, in the initial stages of use of symbolic manipulators in education, Hodgson observed:

In spite of the fact that symbolic manipulation systems are now widely available, they seem to have had little effect on the actual teaching of mathematics in the classroom.   (Hodgson, 1987, p. 59)

He quoted a report of Char *et al* (1986) on the experiences of using the symbolic system *Maple* in an undergraduate course in which students were given free access to the symbolic manipulator to experiment on their own or to do voluntary symbolic problems which they could elect to count for credit. He noted a "somewhat limited acceptance of Maple by the students":

While many explanations can be put forward for such a reaction (little free time, no immediate payoff, weaknesses of the symbolic calculator for certain types of problems, absence of numerical or graphical interface, lack of user-friendliness), it is clear that the crux of the problem concerns the full integration of the symbolic system to the course in such a way that it does not remain just an extra activity. This calls for a revision of the curriculum, identifying which topics should be emphasized, de-emphasized or even eliminated, and for the development of appropriate instruction materials.

(ibid.)

Subsequent developments have seen *Maple* extended to include both numerical and graphical facilities and improved radically in user-friendliness. Yet there is an underlying reason why there may be a major problem with symbolic manipulators in mathematical education which is more than a question of interface, available facilities, and the need for integration in the curriculum. A symbol manipulator is a *tool* – a very powerful tool – but any tool can only be used to its fullest capabilities by those who know how to use it. The situation is parallel to the use of simple calculators: they do not teach a child how to add (or divide), but they are useful tools for adding or dividing when one knows what arithmetic is all about. Once one knows how to cope with small numbers, perhaps the calculator can be used to investigate facts with much larger numbers. Likewise, symbolic manipulators are likely to prove more useful – as they have proved useful in mathematical research – once the student has progressed to the stage of knowing what the tool is being used for.

The later generation of symbolic manipulators, particularly *Mathematica*, have made a step in helping the user come to terms with the nature of the concepts by including word-processing facilities as well as symbol manipulation. This allows the development of teaching material in the form of electronic notebooks, in which symbols present may be manipulated or edited at will by the user. In this way it is possible to introduce the user to new concepts in a cybernetic environment which responds to the users needs in manipulating the symbols which appear. It promises to be an exciting development which has been met with more enthusiasm than the environment which requires the user to type in the complete command in the idiosyncratic syntax of the particular manipulator. Here words can tell the user the meaning of a command and the user may just select it and instruct the computer to carry it out. However, our experience in all the earlier chapters tells us to beware of the simple solution. It is likely to contain seeds for misconceptions and cognitive conflict. In order that students can re-construct their knowledge faced with the radical new concepts of advanced mathematics, they need to gain experience of how the ideas work and actively reflect on the cognitive changes required to integrate this new knowledge into a more appropriate mental structure. Two thousand years ago Euclid is reported to have told Ptolemy that there is no Royal Road to Geometry, given the nature of the human animal, even in collaboration with the computer, we should not be deluded into believing that the computer will provide an entirely smooth path to mathematical knowledge.

Having a computer to perform the algorithms, even to show how those algorithms work is one thing, being able to cope with these concepts meaningfully is another. Some symbolic manipulators include facilities to allow the user to step through the manipulation, seeing what is done at each stage. This can be very helpful to the student who is trying to learn how to reproduce the algorithm, but knowing how to differentiate symbolically is very different from knowing what the derivative means. Likewise, knowing routines for solving differential equations symbolically by reversing this symbolic differentiation process is a very different process from being able to visualize a solution or a family of solutions. What may help to broaden the student's understanding is to set the use of the symbolic manipulator in an appropriate conceptual environment.

5. CONCEPTUAL DEVELOPMENT USING A COMPUTER

Heid (1985,1988) spent the first twelve weeks of a fifteen week applied calculus course studying fundamental concepts using graphic and symbol-manipulation software to perform routine calculations whilst she focussed the students on the underlying concepts. Only in the last three weeks did they practice any routine algorithms for differentiation and integration. She found the learning of fundamental concepts was greatly improved in the experimental class:

> Students showed deep and broad understanding of course concepts and performed almost as well on a final exam of routine skills as a group who had studied the skills for the entire fifteen weeks.
>
> (Heid, 1985, p. 2)

In the classes the experimental students were encouraged to use a large variety of concept representations and to reason with them, for instance using computer generated graphs and tables of values to solve real world problems and make conclusions about applications:

> One student, for example, located the sales level for maximum profit by finding the x-value for the greatest vertical difference between the revenue and cost curves. Another formulated consumers' surplus as the sum of the areas of rectangles without the typical first translation to a Riemann sum formula. A third gave a new integration formula for the area between curves by conjuring up an alternative geometric explanation and translating it directly into a statement about integrals. Reasoning in non-algebraic modes of representation characterized concept development in experimental classes.
>
> (Heid, 1988, p. 10)

By encouraging the students to think for themselves and to construct their own ways of handling the concepts, it became apparent that they had integrated the ideas into their own knowledge structure:

> ... when the students realized that they had made misstatements about concepts... on many of these occasions, on their own initiative, the students in the experimental classes reconstructed facts ... by returning to basic principles. When [they] spoke about limits, functions, derivatives and Riemann sums, the wording was often clearly their own.
>
> (*ibid*, p. 15, 16)

In contrast:

> When the students in the comparison class verbalized that they had made erroneous statements about concepts, there was no evidence of attempts to reason from basic principles. They often alluded to having been taught the relevant material but being unable to recall what had been said in class.
>
> (*ibid*, p. 16)

Thus Heid's research shows clear evidence of the value of giving *meaning* to the basic concepts, even before the students have had any extended practice with the algorithmic techniques.

6. THE COMPUTER AS AN ENVIRONMENT FOR EXPLORATION OF FUNDAMENTAL IDEAS

In her research, Heid used existing software for graphs and symbolic manipulation to build conceptual insights. This software is built on *mathematical* principles: to draw graphs, to carry out mathematical processes, and so on. Another possibility is to design software which uses a combination of mathematical and *cognitive* principles – building on what students already know in a way which is consistent with their cognitive development.

Students meeting advanced mathematical concepts such as infinite processes, limits, continuity and differentiability for the first time are known to have serious cognitive difficulties (see chapters 10, 11). The mathematics educator, with a knowledge of both the mathematics and the cognitive development, can play a fundamental role by identifying powerful ideas in the theory that can be presented in a meaningful way to the students at their current point in development, yet play a fundamental role throughout the theory. To illustrate this we return to the cognitive approach to the calculus illustrated in chapter 11 and concentrate on the computer environment which it uses.

Graphic Calculus (Tall, 1986, Tall *et al*, 1990) was conceived as an example of software designed to provide students with a cognitive approach to the calculus and differential equations. Because of students' known conceptual difficulties in understanding the limit concept, it was decided to found the approach on the notion of *local straightness*. Here the possibility of computer magnification of graphs allows the limiting process to be implicit in the computer magnification, rather than explicit in the limit concept. Students therefore begin the calculus by exploring the magnification of graphs of functions of one variable. They can see that most of the familiar graphs (polynomials, trigonometric, exponential, logarithmic and their combinations) are all locally straight, but some, such as $f(x)=|\sin x|$ have points where left and right gradients differ. They can be guided to look at graphs such as $f(x)=x\sin(1/x)$ (with $f(0)=0$) which oscillates so wildly that it never looks straight at the origin, whilst $f(x)=(x+|x|)\sin(1/x)$ looks straight to the left from the origin, but not to the right. Other functions are available for exploration, including fractal functions that are so wrinkled that they never look straight under magnification, giving students mental images of differentiability and various ways in which non-differentiability may arise. Thus the local straightness of differentiable functions, and non-straightness of non-differentiable functions allows the student to gain a fundamental insight into the notion of differentiability from the very beginning, instead of founding their understanding on simpler ideas concerned only with polynomials.

Local straightness also links naturally to the ideas of differential equations (building locally straight curves, knowing their gradient) and to the general study of differentiable manifolds (locally flat substructures of higher dimensional spaces). The idea is also enshrined in non-standard analysis (e.g. Keisler, 1976) where it is proved that under an infinite magnification (the standard part of) an infinitesimal portion of a graph is precisely straight.

As discussed in chapter 11, a student with the mental ability to view a small part of a displayed graph and to see its gradient, can then conceptualize the numerical gradient $\frac{f(x+h)-f(x)}{h}$ for variable x and fixed h. By investigating the numerical gradient in simple cases using the computer, it is found that students can conjecture the formula for the stabilized numerical gradient, which is the derivative, before they have the ability to derive the formula algebraically from first principles (Tall, 1986).

Furthermore, the pictorial idea can lead to the notion of a differential. If a graph is locally straight, then a small portion of the tangent at a given point (x,y) will closely approximate the curve. Denoting the components of the tangent vector by dx,dy then $f'(x)=\dfrac{dy}{dx}$ and visually, one can see that the point $(x+dx, y+dy)$ is on the tangent, closely approximating the curve when dx is small.

This leads naturally into the notion of a first order differential equation

$$\frac{dy}{dx} = F(x,y)$$

where the gradient at any point (x,y) is given as $F(x,y)$. The *Solution Sketcher* (Tall, 1989) allows the user to specify a first order differential equation, then move a pointer round a screen window representing the (x,y) plane, drawing a small line-segment through (x,y) with gradient $F(x,y)$. By a simple key stroke the line-segment may be left as a permanent mark, and successive segments may be placed end to end to *construct* an approximate solution to the differential equation. Thus the student can gain a physical idea of what the solution of a first order differential equation actually *means*.

It is a simple matter to show that a higher order differential equation

$$\frac{d^2x}{dt^2} = F(t,x,\frac{dx}{dt})$$

can be written as two simultaneous differential equations by the substitution $v=dx/dt$ to get

$$\frac{dv}{dt} = F(t,x,v), \frac{dx}{dt} = v.$$

This too has a (locally straight) solution in (t,x,v) space with tangent direction given by

$$(dt, dx, dv) = (dt, v\ dt, F(t,x,v)\ dt)$$

which is in the direction $(1, v, F(t,x,v))$. Thus the simple idea that a solution "follows the gradient direction" is true not only for first order differential equations, but for higher order (simultaneous) differential equations in a suitable solution space.

Hubbard & West (1985) developed a computer graphics approach to differential equations. They found that, without computer graphics, students had difficulty appreciating the notions of existence and uniqueness of solutions. When so much of their work had involved routine symbolic manipulation to produce an answer, many students found it difficult to comprehend how a solution could exist if it could not be expressed as a familiar formula. The computer graphics helped them to see existence as the ability to draw a solution – a solution that existed visually even though they were unable to provide a formula for it. This links closely to the formal theory – the solution exists, and is unique, provided that the differential equation properly specifies a direction to follow at each point. Solutions fail to exist where the differential equation fails to specify a unique direction.

The fact that there are symbolic differential equations which lack symbolic solutions shows the need to incorporate numerical and graphic representations with symbolic methods. A computer is able to process a vast amount of numerical data and to present it in graphical form. Even where the symbolic methods are available, they may need geometric interpretation. Tall (1986c) quotes the following example from a national examination in the U.K.:

$$y\frac{dy}{dx} \sec 2x = 1-y^2.$$

It is easily solved by separating the variables:

$$\frac{y}{1-y^2} \, dy = \cos 2x \, dx, \qquad\qquad (*)$$

and may be integrated to give:

$$-\frac{1}{2} \ln |1-y^2| = \frac{1}{2} \sin 2x + c.$$

But what does this *mean*? Regarding (*) as specifying the direction of the tangent vector (dx,dy), to the solution curve through any point (x,y) enables a "direction field" of short line segments to be drawn in the appropriate directions through an array of points in the plane (figure 33).

It can be seen that some solutions are closed loops whilst others may be conceived as functions in the form y=f(x). The symbolic solution is in this case of little value without a graphical representation of its meaning, whilst the graphical interpretation alone lacks the precision of the symbolism. It therefore needs the complementary power of visualization and symbolic manipulation to give a deeper mathematical insight.

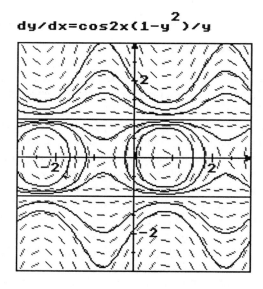

Figure 33 : Solutions to a first order differential equation

7. PROGRAMMING

In recent years moves have been made to introduce student programming into mathematics courses. Initially this tended to be in the form of enhancing ready existing mathematics courses by introducing computers and calculators to carry out numerical algorithms and perhaps represent the results graphically. It has met with mixed success. With younger children there is considerable evidence that if programming is simply attached to a course without any thought about conceptual integration, then there is no reason to expect an improvement in conceptualization of the course content (Menis *et al*, 1980; Cheshire, 1981). Some research projects have shown that when programming is introduced as an extra into the traditional curriculum it may reduce the time spent on traditional skills, causing a lower level of performance in them (Reding, 1981; Robitaille *et al*, 1977). However, Thomas & Tall (1988) found that teaching algebraic concepts in a module including programming at first gave the usual initial losses in traditional skills to balance gains in conceptual understanding, but after a brief review of skills at a later date, this was changed into a gain on both skills and concepts.

At university level, Simons (1986) reported on the use of hand-held computers to be programmed in BASIC to supplement the traditional teaching of calculus in these terms:

> ... the introduction of a personal computer into a course of this nature, whilst enhancing teaching and presentation in many areas, raises profound problems.   (Simons, 1986, p. 552)

There were evident gains in the immediate usefulness of the work, but a substantial number of staff, long experienced in mathematics teaching yet new to the computer and numerical analysis, did not like the course. Simons suggests that the aversion displayed by some members of staff lies in the feeling of uncertainty in applying a numerical method:

> The traditional mathematician ... is clearly aware that for every numerical method a function exists for which the method produces a wrong answer. ... The statement that nothing is believed until it is proved is the starting point for teaching mathematics and introducing the computer forces the teacher away from this starting point.   (*ibid*, p. 552)

A recurring observation is the difficulty experienced by teachers, both at university and in school, to come to terms with the new technology. We are at present in the throes of a paradigmatic upheaval and cultural forces operate to preserve what is known and comfortable, and to resist new ideas until they are proven better beyond doubt.

On the other hand, there is also evidence that when programming is used for conceptual purposes, such as solving problems where the programming parallels the underlying mathematical processes, or using computer activities to foster specific mental constructions that can lead to mathematical understanding, then there is a much higher level of success.

Several universities in the U.K. now include mathematical problem-solving through programming – usually in structured BASIC – as an element of the undergraduate mathematics course. The problem-solving often requires program construction to give numerical or graphical data and experience shows that the students gain considerably from the task.

Various programming languages are becoming available which are almost certainly more appropriate for mathematics than BASIC. Some are specifically designed to make concepts in mathematics easy to program. For example, *Mathematica*, as well as providing a symbolic manipulation system within a word-processing package that will draw graphs, also gives a complete programming system that allows a powerful blend of functional and structural programming constructs. Such developments within a multi-purpose computer environment are likely to prove of increasing use in advanced mathematical thinking in the future. It should be noted, however, that the principal aim of the programming system of *Mathematica* is predominantly for *doing* mathematics, rather than *learning* mathematics. It is therefore better designed for the expert than the novice.

A language specifically designed for mathematics learning is ISETL (Interactive SET Language). Dubinsky and his colleagues have found that having students make certain constructions in the ISETL can lead to their making parallel mathematical constructions in their minds and thereby come to understand various mathematical concepts (Ayers *et al*, 1987; Dubinsky, 1986, 1990a, 1990b; Dubinsky *et al*, 1988; Dubinsky *et al*, 1989; Dubinsky & Schwingendorf, 1990a, 1990b). The specific use of computers in this work is driven by the theoretical analysis laid out in Chapter 7 and a brief description of the language is given in an appendix to this chapter.

These experiences, both positive and negative, tell us that the issue in using programming to help students learn mathematical concepts is not whether it should be done, nor is it the particular language that is used. The main consideration is how the instructional treatment uses the language through the design of the programming tasks for the students.

Although the nature of the computer language is not the primary consideration, it is an important one. The inconvenience of working with Fortran or Pascal syntax introduces difficulties for students and teachers that have nothing to do with mathematical issues. The same is true to a lesser extent of LISP, APL and PROLOG. BASIC is easier to use and is adequate for numerical algorithms and representing numerical data in a graphical form, but it is inappropriate for arithmetic with large integers, for symbol manipulation and for most higher-level mathematical thinking. LISP is particularly powerful for symbol manipulation and LOGO is almost as good (for the purposes of mathematics) with much less syntactic overhead. APL makes working with vectors and matrices especially easy while PROLOG is designed for programming systems of complex logical inferences. ISETL supports most of the standard mathematical constructs with a syntax very close to mathematical notation. It is the only one of these languages that treats functions as data.

Graphics have often only been added to languages at a later stage. The omission of graphics from the first version of BASIC led to a proliferation of different dialects. Given the acclaim for turtle graphics, it is a salutary experience to realize that these were almost an afterthought in the initial specification of LOGO. ISETL was also originally designed without graphics which were added later.

To ask which kind of programming language is most beneficial to help students learn mathematics, one must first ask what it is one is trying to teach and how:

Is mathematics a bag of tricks that may be useful to later life? Is mathematics taught because it is an important part of our culture, or because it helps young people to teach logically and abstractly? These are questions for mathematics teachers. In the long run, computer software can be adjusted to their requirements.
                                                                          (Grogono, 1989)

Grogono shows how different kinds of languages may be used to model different kinds of thinking processes. The question is equally applicable at more advanced levels of mathematics. If its answer is that one wishes to encourage students to think mathematically about mathematical concepts, then a computer language is required that supports these requirements.

8. THE FUTURE

Thus we see the computer already proving a powerful tool in advanced mathematical thinking, both in mathematical research and in mathematics education at the higher levels. The empirical evidence shows that it proves more successful in the educational process when it is used to enhance meaning, either through programming in a language embodying the mathematical processes or through the use of computer environments for exploration and construction of concepts.

Computers are likely to prove a profound influence over the next N years, where the reader may care to estimate the value of N. It is possible, but it may not be meaningful, to speculate on the changes that new technology will bring. Already the promise of parallel processing may bring new possibilities, for instance in the simultaneous processing of several different representations. Intelligent tutoring systems currently seem to promise more than they deliver, but it is conceivable that new techniques may bring greater success. Already we have video discs carrying large amounts of information for the user to explore in new and unforseen ways.

However, it is our belief that mathematics is not a spectator sport, and that advanced mathematical thinking will continue to blossom through the constructive actions of the human mind, albeit complemented by the enormous processing power of the computer.

APPENDIX TO CHAPTER 14

ISETL : A COMPUTER LANGUAGE
FOR ADVANCED MATHEMATICAL THINKING

ISETL is a computer language which has been designed and used to foster mathematical thinking at advanced levels. The language and its use will be indicated by giving some examples of actual code along with indications of how this relates to some specifics of the constructivist analysis given in Chapter 7. We will use terminology such as process, object, interiorize, encapsulate, coordinate, and reverse which are explained fully in that chapter.

The interactive set language ISETL is designed to implement many mathematical constructions in ordinary mathematical language. Sets can be listed in the usual way within braces {}, either as a list of elements separated by commas, or as a set defined by a property. Square brackets [] denote sequences, and the notation [a..b] for integers a,b denote all the integers from a to b.

The following line entered into ISETL:

$$P := \{x : x \text{ in } [2..1000] \mid \text{not exists } y \text{ in } [2..(x{-}1)] \mid x \text{ mod } y = 0 \}$$

assigns to **P** the set of numbers x between 2 and 1000 which do not have a smaller factor **y** – in other words **P** is the set of primes less than 1000.

In full generality a set in ISETL can be specified as:

$$\{ \text{ expr} : x,y, \dots \text{ in } S, u,v, \dots \text{ in } T, \dots \mid \text{condition} \mid \text{condition} \dots\}$$

where **expr** is an expression, generally involving variables **x**, **y**, **u**, **v**, etc whose domains are previously constructed sets **S**, **T**, ... and each condition is an expression whose value is true or false. It is important for the student to think about how the computer might handle this construction: by iterating the variables through their domains, and for each value to evaluate the conditions and, if it is true, placing the expression in the set.

The assumption made by those who use this language in education is that by writing such code the student will interiorize the process of forming this set.

A set is not only a process of formation, it is an object with its own existence; for instance, it has a cardinality operator, it can be itself a member of a set, etc. One way to check that someone has an understanding of the process is to ask her or him to calculate the number of elements in a set such as

$$\{1{+}2, \{1..4\}, \text{``cat''}, \{1,2,3\}, \{\{\text{``house''}, \text{``dog''}, 3\}, 3\}\}$$

(in this case it is 5). ISETL does this with a single operation. Thus,

$$\#(\{1{+}2, \{1..4\}, \text{``cat''}, \{1,2,3\}, \{\{\text{``house''}, \text{``dog''}, 3\}, 3\}\});$$

returns the value 5.

Again there is an assumption that if you write code that applies operators, then you will tend to think of that to which an operator applies as an object. In this way, it is considered that students will come to encapsulate the process of set formation and think of the resulting set as an object.

A function can be represented in ISETL as a dynamic process which transforms elements in one set to elements in another. For instance:

```
F := func(k);
        return %+[i**2 : i in [2,4..k]]
    end;
```

defines F as a function of k and returns the sum (denoted by %+) of the squares of all even numbers between 2 and k.

An important effect of writing procedures that express mathematical actions is that, in the sense of Chapter 7, the students tend to interiorize these actions and construct mental processes that contribute to their understanding the underlying concepts.

As we pointed out in Chapter 7, it is important to encapsulate functions that are understood as processes and think of them as objects. The best way to achieve this is to operate on functions and/or make new ones. This is possible in ISETL because a function is treated as data. It is possible to form sets of functions, have functions as parameters to other functions and also to have a func construct and return a function. Consider the following example.

```
co := func(f,g);
    return func(x);
        return f(g(x));
        end;
    end;
```

co is an operation which will take two representations of functions, say f1 and g1 and return a representation of their composition. The composite function co(f1,f2) may also be written using infix notation as (f1 .co f2). Assuming that f1 and f2 represent functions and the value of **expr** is in the right set, the computer will accept

(f1 .co g1) (expr);

and return the value of

f1(g1(expr)).

A powerful way to use this idea is to have students construct **co** and use it, preferably to solve problems of interest to them. The student will tend to have a number of important experiences as a result of constructing **co**. First, it is necessary to think of functions as objects in order to imagine applying some process to two functions. Then these two objects must be unpacked to reveal their processes which can be coordinated by linking them sequentially. The resulting process is then converted back to an object by the three lines beginning with **return func(x);**. This code, which has the effect of returning a representation of a function whose domain variable will be denoted by **x**, is very difficult for students and having them struggle to construct it in order to solve a problem can have a profound positive effect on their conceptualization of functions.

A second way of representing functions in ISETL, which corresponds to one way that mathematicians think of functions is to list the ordered pairs, for instance

H := {[x,x**2] : x in P};

assigns to the variable H the set of ordered pairs [x,x**2], where x is a prime less than 1000 and x**2 denotes x^2.

Within ISETL a set of ordered pair works like a function, so an expression such as

H(3);H(7);H(4);

will print on the screen the values **9, 49** and **om**, the last symbol being the sign that H(4) is not defined because 4 is not in P.

In a sense, this reverses the mental excursion. If a function is constructed as a func which is then operated on, one is influencing students to think about a function first as a process, then as an object. A set of ordered pairs, on the other hand, is most likely to be considered to be an object, especially if previous study of the language has treated sets in this way. Having students write such code and then do evaluations tends to have them think first of a function as an object, and then as a process. Clearly, students should experience both excursions and see them as two aspects of the same notion. The fact that ISETL will treat sets of ordered pairs and funcs in many similar ways (for example, **co** will work just as well if its inputs are sets of ordered pairs rather than funcs, or even a mixture) helps students unify their thoughts about the two points of view.

An example of the inputs to a function being a combination of functions and numbers is the following func to calculate a Riemann sum for the function f from a to b using n equal width strips whose height is the left endpoint of each subinterval:

```
RiemLeft := func(f,a,b,n);
    x := [a+((b–a)/n)*(i–1) : i in [1..n+1]];
    return %+[f(x(i))*(x(i+1)–x(i)) : i in [1..n]];
end;
```

Students can also encapsulate the notion of integration as a function operating on other functions by defining:

```
Int := func(f,a,n);
    return func(x,a,n);
        return RiemLeft(f,a,x,n);
        end;
    end;
```

Here **Int(f,a,n)** represents a function of f, a and n where **Int(f,a,n)(x)** gives the Riemann sum for f from a to x using n equal steps.

ISETL is also ideal for other mathematical concepts and the benefits to learning can also be delineated in terms of the general theory presented in Chapter 7. We mention briefly a few additional things one can do in this language and how they relate to understanding mathematical concepts.

For instance, it is helpful for students to write programs to construct the truth table for a given expression. With the first order calculus there is again the dichotomy and synthesis of thinking of a logical expression as a process and as an object. Thus, in an expression such as

$$(P \wedge Q) \Rightarrow ((-Q) \vee (P \vee R))$$

the expression $(P \wedge Q)$ can represent, in the mind of the student, a process consisting of putting together P and Q and evaluating the truth or falsity for various values of the variables. But in order to combine $(P \wedge Q)$ with the rest of the expression, it must be treated as an object.

Once boolean expressions (having the value true or false) are considered as objects, they can be collected as elements in a set. This is a critical step in the transition to the second order predicate calculus in which quantification is involved. In order to interpret the logical statement

$$\exists x \in S \ni P(x)$$

one has to imagine a set of propositions indexed by x. The existential operator is performed by iterating x through the domain S, evaluating the proposition valued function P at x and, if once the result is true, declaring success and going home. This is exactly what the computer does when given the ISETL command

exists x in S | P(x)

and thinking about the ISETL procedure helps the student think about the corresponding mathematical process. Beginning with a function P of two variables and applying two quantifiers (generally one existential and one universal) leads to a second order quantification. Writing the code helps the student to coordinate two instances of the quantification process and make the appropriate mental construction.

Formal definitions of mathematical structures are straightforward to implement (for finite sets) in ISETL. For instance, if G is a finite set with binary operation **op**, then the following ISETL func tests whether it is a group:

```
grouper := func(G,op);
    return (forall x,y in G | x .op y in G)
        and (forall x,y,z in G | (x .op y) .op z) = (x .op (y .op z))
        and (exists e in G | (forall x in G | x .op e = x))
        and (forall x in G | (exists y in G | x .op y = e));
    end;
```

Notice how closely this code resembles the formal definition of a group. It also fosters the psychological constructions necessary to understand the group axioms. There are several instances of processes and objects here as well as coordination of two processes. In addition, the axiom for inverses requires a reversal of the process which arose in the axiom for identity.

It turns out that, whether or not the students succeed in writing such a func, once they have it and understand it, they can write funcs to test for subgroups and even normal subgroups. Then, it is very effective to have them construct the set of cosets, define the appropriate binary operation and use grouper to decide whether it gives a group. This can be carried at least up to the fundamental theorem of homomorphisms.

EPILOGUE

REFLECTIONS

DAVID TALL

The production of this book is a first stage in a journey which sixteen authors and a wider group of co-workers in Advanced Mathematical Thinking have shared. It is pertinent, given the nature of the thinking processes that we have unfolded, to reflect upon what we have done with the spiral of conceptual development in mind. First one begins with a problem which may not be well-defined. Then one uses what tools are available to attack the problem as it progressively becomes clearer, with all the false starts and hard-won minor advances that are inevitable ingredients of the struggle. And now there is a calm after the storm to reflect, to see what gains have been made and what remains to be done.

It would be good to be able to look back on the definitive book on *Advanced Mathematical Thinking*, with all the resolutions of all the problems that occur and a coherent theory that explains what it is and how to help others achieve it. This task is not yet complete, certainly not in a definition-theorem-application format that a mathematician might require of a theory. What has been done is to set out on a journey, on which the reader has been encouraged to participate, to consider the way in which advanced mathematical thinking functions, to understand what makes some thinkers successful and to help others on their journey to greater success. "The journey is the reward". And at this time we can look back on the pathways we have taken to see what problems have been well-formulated and what solutions have appeared as we move on to the next stage of the journey.

For me, as editor, it has been a fascinating study to see the development of various parts of a theory, to see consonances and dissonances, some of which have been resolved whilst others remain suspended in the ether. At the beginning of the journey I saw through a glass darkly. I have yet to see face to face.

But now there are clearer avenues to follow, beginning with a more focussed picture of the nature of the advanced mathematical thinking and moving towards pertinent questions and partial answers. First we must highlight the different ways in which individual mathematicians may think successfully. In particular, the need for all of us, successful in our various ways, to give space to others to help them use their own particular talents to build up their mathematical thinking processes. Then there is the realization of the thorny nature of the full path of mathematical thinking, so much more demanding and rewarding than the undoubted aesthetic beauty of the final edifice of formal definition, theorem and proof.

It is clear that the formal presentation of material to students in university mathematics courses – including mathematics majors, but even more for those who take mathematics as a service subject – involves conceptual obstacles that make the pathway very difficult for them to travel successfully. And the changes in technology, that render routine tasks less needful of labour, suggest that the time for turning out students whose major achievement is in reproducing algorithms in appropriate circumstances is fast passing and such an approach needs to move to one which attempts to develop much more productive thinking.

It is therefore no longer viable, if indeed it ever was, to lay the burden of failure of our students on their supposed stupidity, when now the reasons behind their difficulties may be seen to be in part to be due to the epistemological nature of mathematics and in part to misconceptions by mathematicians of how students learn. We often teach certain skills because we know that these will bring visible, albeit limited, success, but we now know, somewhat furtively, that the acquiring of those skills may develop concept imagery that contains the seeds of future conflict. We have evidence that a formal approach, which appeals to the sophisticated expert may be cognitively totally inappropriate for the naïve learner and demands new forms of teaching to pass through the transition from elementary mathematics to a point where the economy and structure of modern mathematics is seen as a meaningful goal.

It seems incredible that our list of references is largely dominated by papers written in the last decade with only a few honourable exceptions before the early eighties. What has emerged from a meeting of individuals over a five year period, to reflect on this newly developing area of concern, is a clearer understanding of the full cycle of mathematical thinking: the need to begin with conjectures and debate, the need to construct meaning, the need to reflect on formal definitions to construct the abstract object whose properties are those, and only those, which can be deduced from the definition. Advanced mathematics, *by its very nature*, includes concepts which are subtly at variance with naïve experience. Such ideas require an immense personal re-construction to build the cognitive apparatus to handle them effectively. It involves a struggle which virtually every author in this book, both severally and individually, sees in terms of a reflection on personal knowledge and a direct confrontation with the inevitable conflicts which require resolution and reconstruction.

College professors see this conflict daily in individual students as they struggle to come to terms with new ideas. In the past they have often tried to help by providing clearer lectures, making the transitions as simple as possible, presenting the ideas in a way which reduces the strain. This may even lead to the successful professor being lauded by his or her students for the clarity of their exposition, but the acid test is *what do the students learn*? And this needs to be assessed in a wider sense than just which algorithms closely related to their course they can carry out, or which definitions and proofs can they correctly reproduce.

Our cognitive studies have shown the manifold differences between the formal definitions of concepts and the images we use in our minds to work with these concepts. They show how the complexity of the subject demands a "chunking" of information in an efficient way so that it can be easily handled, and this is linked to the appropriate use of symbolism for a given context and the appropriate meaning which the individual links to that symbolism.

We have seen a divergence between the visualizers and verbalizers amongst us, just as there appears to be a time-honoured difference between the mental processes of the mathematical giants of the past. In recent months, as I have interacted with the various authors in an attempt to either come to an agreement or to hone our differences into explicit focus, I have been privileged to gain some additional insights.

It is clear that mathematics without process to give results is of little value, in other words, visualizing an idea without being able to bring it to fruition is virtually useless. I emphasize this fact even though a major thrust of my work is in the use of visualization. On the other hand, simply to be able to carry out procedures in a narrow way, without being able to see the overall connections, is also grossly limiting. For me this has led to a belief in a versatile form of thinking which complements the procedural with the global overview. However, we have evidence of mathematicians, such as Hermite, steeped in the logic of their subject who develop a powerful intuition of the processes and their symbolism in such a way as to render visualization – for them – redundant. We also have evidence of successful students (such as the case of Terence Tao, Clements, 1984) who vastly prefer the power of logical deduction. We therefore need to cater for different types of minds.

Recently, however, in a very different context, I was able to obtain an insight which may prove helpful in this apparent dichotomy. Mathematics – according to the Oxford Dictionary – is said to be "the Science of Space and Number". In recent months I have been reflecting on the fundamental differences between these two different forms of mathematics and the manner in which they develop cognitively. Space, through the study of geometry, begins with gestalts – "that is a triangle", "this is a straight line", "that there is a rectangle" and "this is a square". The child learns to recognize these visual gestalts from examples and non-examples. "Yes, that is a square, but don't think it is not a rectangle, because a square is a special kind of rectangle". Through exploration and interaction with others, the child learns to discriminate between these various gestalts and to isolate some of their properties: "a rectangle has four right-angles and opposite pairs of sides equal", "a square has four right-angles and all four sides equal" and to begin to see relationships "an isosceles triangle has two equal angles and two equal sides". From here the relationships begin to build into deductions "*if* a figure is a square, *then* it is a rectangle", "*if* a triangle has two equal sides, *then* it has two equal angles", definitions begin to be isolated and, finally, these can be formulated in an axiomatic way to give the framework for logical deduction. Indeed, what I have just described in outline was formulated about the development of geometry more clearly as a hierarchy over thirty years ago by Van Hiele (1959).

Number on the other hand is a very different animal. It begins with imitation of the number names recited in sequence, "one, two, three, ...", perhaps imperfectly at first, "... four, five, nine, seven, ...", then with more confidence, until the routine of pointing at objects and reciting the number names in proper sequence leads to the concept of counting. This is an encapsulation. The *process* of counting leads to the *concept* of number. By various further strategies of process, "counting all" of two sets (a coordination of two processes), or "counting on" (combining the concept of number of the first set with the process of counting the second) leads from the process of counting to the concept of "sum". A vital phenomenon occurs here in that the symbolism 4+3 represents to the user both the process of counting and the product of that process, the sum. The rest of the "number" part of mathematics proceeds, in the same way, by encapsulating processes as concepts, *often using the same symbolism for both process and concept*. Thus the process of "repeated addition", "five threes" becomes the concept of "product", "5 times 3" – both written as 5x3. The process of "repeated multiplication" becomes the concept of "power" and so on. Of course this prescription is exactly parallel to the discussion of Dubinsky on reflective abstraction. It is a phenomenon known to Piaget and to many an observant teacher since time began – except that there is an amazing simplicity about what is being done. In the number side of mathematics the mathematician makes progress *by being ambiguous* about notation. (S)he uses the same notation for process and product *deliberately*, so that (s)he can powerfully use whichever is appropriate for a given task. To *calculate* means to use the *process*, to *manipulate* is easier with a single object which involves using the product.

As whole number generalizes to signed integer, the symbols +2 and –7 also have dual roles as process and the product: "shift two units right on the number line", "the integer plus two", "shift seven units left", "the number minus two". The same happens with fractions: "3/4" is both "divide three by four" and the product of the process: "three-quarters". It is the same with trigonometry, where

sine = opposite/hypotenuse

is both an instruction to calculate and a symbolism for the result.

Algebra too exhibits this same dualism of notation where $2+3x$ means both the process of adding two to the product of three and x and also the result of that process.

In chapter 4 Hanna remarks on the irony that in a "discipline touted as precise, the student must develop a tolerance for ambiguity". Instead of being defensive about this state of affairs, it is more appropriate to note that the successful mathematician is the individual who sees the duality of this kind of notation as process and product and who uses the ambiguity in a flexible way. Given the importance of a concept which is both process and product, I find it somewhat amazing that it has no name. So I coined the portmanteau term "*procept*" for a process which is symbolized by the same symbols as the product. It seems that the whole of number and algebra is built on procepts, so a theory of procepts and their use in mathematics has a vast potential domain of application.

Yet space and geometry are different. They seem to be built on gestalts whose properties are only slowly teased out and put into coherent relationships, then definitions and deductions.

There are therefore (at least) two different kinds of mathematics. One builds from gestalts, through identification of properties and their coherence, on to definition and deduction at advanced levels of mathematical thinking. The other continually encapsulates processes as concepts, to build up arithmetic, then generalizes these ideas in algebra before formalizing them as definitions and deductive theorems in the advanced mathematics of abstract algebra.

If we look at the discussion of Vinner in chapter 5, we find his theories originally began with *geometry*, and his examples include "car", "table", "house", "green", "nice", etc. *None of these are procepts*. However, if we look at the discussion of Dubinsky in chapter 7, we find his examples include "commutativity of addition", "number", "trajectory" (as a coordination of successive displacements), "see-saw" (as the balancing of two objects), "multiplication", "fluid levels" (as a 'variation of variations'). All these are *processes* which become encapsulated as *concepts*. As they stand, they are not all procepts within the narrow meaning of the term just defined. However, they all involve manipulation of quantities, or balancing of quantities, or variation of quantities, and this in turn involves number, which brings us back to proceptual ideas where symbolism is used both to represent a process of manipulation and the result of that process.

The fascinating thing is that, by the time we reach the level of formalism in advanced mathematics, these two different strands move to a similar formulation: the *definition* of concepts and the *deduction* of properties of those concepts.

I believe that the major catastrophe of the new mathematics movement was due to the unproven assumption that "if only the concepts are properly defined, then everything will be OK". The need for clear definitions and deductions caused mathematicians to be covert about the power of their ambiguous use of procepts. This move served our students badly because it failed to acknowledge the methods of the working mathematician. The power in mathematics is not given through unique and precise meaning to symbolism – "a function is a set of ordered pairs such that ..." – but through a *duality* which gains *flexibility* through *ambiguity*[1] – a function is both a *process* (to be able to calculate) and a *concept* (which can be manipulated). It is as simple as that. We cheated our students because we did not tell the truth about the way mathematics works, possibly because we sought the Holy Grail of mathematical precision, possibly because we rarely reflected on, and therefore never realized, the true ways in which mathematicians operate.

The evidence which we are collecting with a wide range of ability of much younger children is that the most able naturally use this flexibility (Gray, 1991). In arithmetic they soon learn a few facts then, when they are faced with a new arithmetic problem, they are often able to relate it to one they know and derive new facts from old. The more able therefore have *a built-in knowledge generator* that develops new arithmetical knowledge from old. Once they grasp this, they realize that they do not need to remember so much because they can soon derive what they want to know. They have a flexible *proceptual* knowledge in which number problems such as 4+5 can be decomposed as the process 4+5, which might be seen as 4+4+1 and (if they know 4+4=8) can be reassembled as 8+1=9. Thus the procept 4+5 is decomposed into process and parts of this are recomposed back to derive the concept, or result, 4+5=9.

Meanwhile the lower ability children remember few facts and continue to use the process of counting to add numbers together. If asked 8+4, they faithfully count on four to get "nine, ten, eleven, twelve" but this is rarely remember as a known fact and, instead of having a knowledge generator, *they have an unencapsulated process which produces answers which are not manipulable objects.* Thus there grows a "proceptual divide" between the more able, using proceptual flexibility, and the less able, locked in process.

The same proceptual divide occurs with algebra. The child who sees algebraic notation only as process, is faced with a nightmare, for how can (s)he conceive of 2+3x as a process when, without knowing x, it is a process which cannot be carried out. And if x is known, why is it necessary to use algebra anyway? Only the child who can give meaning to the symbolism as a conceptual entity can begin to manipulate more complex expressions meaningfully in the sense of Harel and Kaput in chapter 6.

This same division between those who conceptualize process as product and those locked in process occurs again at higher levels. The limit concept $\lim\limits_{n \to \infty} a_n$ is again a *procept*. The same notation represents both the *process* of tending to the limit, and also the *value* of the limit. But this phenomenon is very different from procepts met in elementary mathematics. There the process could be used to *calculate* the product. Now we have the phenomenon that Cornu identified as an obstacle in chapter 10: understanding the dynamics of the process does not lead directly to the calculation of the limit. Instead indirect alternative methods of computation must be devised.

Just as with arithmetic, the theory of limits has a structure for devising new facts from old. But in arithmetic the new facts are derived from old using the calculation processes of

---

[1] I am grateful to my colleague Eddie Gray for this phrase, which comes the title of a joint paper (Gray & Tall, 1991) based on his work with the number processes of younger children.

arithmetic and the new facts have the same status as the old: they can be calculated by the processes of arithmetic in the same way. In the case of the theory of limits, the "known facts" are one or two "elementary" deductions from the definition: that $\lim_{n \to \infty} 1/n$ is zero, or that a constant function and the identity function are continuous. All three of these "elementary" facts are derived from the definitions in singularly peculiar ways which can cause initial confusion. The fact that $1/n$ tends to zero might be deduced from Archimedes' axiom, or perhaps by some heuristic appeal to the fact that :"I can make $1/n$ smaller than ε by making n bigger than the integer part of $1/\varepsilon$ plus one", both of which are strange ways of asserting $1/n$ gets small as n gets large – the student *knows* that anyway! To establish the fact that a constant function is continuous is just "tell me ε and I will tell you δ, in fact you can take any $\delta > 0$ you like, say $\delta = 1066$". It is a joke that few students have the experience to find funny. The continuity of the identity function is equally enigmatic "OK, take $\delta = \varepsilon$, then, when $|x - x_0| < \delta$, we have $|f(x) - f(x_0)| < \varepsilon$, because $x = f(x)$, can't you see, you dummy?" Unlike arithmetic, once these few "elementary facts" are deduced, few, if any, other such "facts" are calculated directly. Instead the "algebra of limits" is proved, using the coordination of the "unencapsulated definition of the limit" as reported in chapter 10, which is at, or beyond, the zone of competence of most students. The result is that the derived facts are "proved" (any polynomial is "continuous" by an induction argument combining sums and products of constant functions and the identity) yet the actual definition is no longer used because the calculations become horrendous.

Thus it is that the procepts in advanced mathematics work in a totally different and completely enigmatic way compared with the procepts in elementary mathematics. It is no wonder that, faced with this confusion, so many students end up conceiving the limit either as an (unencapsulated) process or in terms of meaningless rote-learned symbol pushing.

Likewise the gestalt geometric concepts work differently in advanced mathematics too. Instead of being "described" and having coherent relationships, they are "defined" and other properties must be "deduced" from the definitions. Again, given the conflict between the elementary ideas where the facts are known and the abstract ideas where they need to be deduced confusion, as discussed by Vinner (chapter 5), is almost inevitable.

So what is the solution? First it should be noted that the chapters of this book nowhere give methods that will produce guarantied success. There is no dispute that, for the most able, a formal presentation may be sufficient to show the structure of the subject which they may appreciate and build into a deductive system. But for the vast majority of students, the way ahead is stony and littered with cognitive obstacles which, if not addressed, will only be isolated in the mind in such a way that they lie there ready to cause conflict in future times – if they do not cause outright confusion already.

The evidence is that students of a wide range of abilities prosper when they can give meaning to the ideas. This does *not* mean that they must always relate the concepts back to some concrete foundation that has physical meaning. Just as the child who counts objects successfully moves on at a very early age to mentally manipulate number symbols in arithmetic, so successive layers of encapsulation of process into procept only need refer to the level of the previous proceptual layer. In fact, once the encapsulation has occurred, the use of the same symbolism causes the process and concept to coalesce into a single level. Thus the so-called hierarchy of concepts, which is an obstacle to learning, becomes, to the successful encapsulator, a single level in which process and concept are dually represented, with the complexity disguised by the simplicity of the symbolism.

The question to be addressed is: if this is the way of success for the more able, what should we do with, or for, or to, the vast majority of our students? The evidence in this book is, that to give them a sense of the full range of advanced mathematical thinking, it is essential to help them reflect on the nature of the concepts and the need for mental reconstruction in an overt and explicit way, and to give them opportunities in which they can learn to conjecture and debate, so that they may participate in mathematical thinking, not just learn to reproduce mathematical thought.

This is not going to be an easy task. What stands against it is, in many cases, fear. Fear of professional mathematicians for the unknown when they leave their neatly planned course structures of theorem-proof-application and give open-ended opportunities for problem-solving. Fear of the increased time that this will take and that they will not "get through the course". Fear that "standards will drop" because students will not be able to exhibit the ability to carry out all the processes that need to be taught in an "honours degree". Fear that they dare not make any changes whilst other institutions maintain the traditional standards.

In recent years the fast changes in society are causing all of the well-established truths to be reassessed. In Britain through the Institute of Physics, university departments of physics have mutually agreed to reduce the content of the three year physics course by one third to give more room for understanding what is actually taught. In mathematics a step in similar direction might not be out of place. It is not necessary to change the whole of the approach in a single step. Given a modest reduction in content, a new flexibility could allow, say, a single course in problem-solving, of a general nature, to be introduced early in the course, to encourage creativity in mathematical thinking, even though it introduced no new content, but compensated in terms of reflection on higher processes. For ten years I have run such a problem-solving course and I know the way it changes students' perceptions of themselves and builds up confidence through success in small things that steadily grow more complex. They learn to *talk* to each other, to *verbalize* mathematics, to *speak* coherently. They even learn to enjoy interchanging information and helping each other, whereas before they had often believed that good students only do mathematics for themselves, on their own.

Given a modest reduction in content, it might be possible to allow time for students to explore their own conjectures in a specific subject area. In my own analysis lectures I regularly set up a problem scenario and leave the students to work in groups to try to solve the problem. "OK, so the intermediate value theorem seems obvious, but suppose you knew f was continuous between a, b and that $f(a)$ and $f(b)$ had opposite signs – how would *you* prove that it is zero in between?" Setting this as homework does not have the same effect as encouraging students to talk together in class time, and the best way to do this is for the instructor to make sure that there is a good topic for investigation and then leave the room. Some of my best teaching occurs when I am somewhere else drinking coffee and getting paid for it! A return to the classroom after an appropriate passage of time may find that the students have not solved the problem, but they often have experiences on which a proof can then be constructed through a mutual dialogue. In this way they learn to participate in the construction of mathematical knowledge rather than just remembering and repeating it.

Viewing the third part of this book – the review of the literature – we see authors adopting very different stances. Robert and Schwarzenberger highlight the difficulties of the transition from school to university. Eisenberg begins with a catalogue of failure in the teaching of the function concept. Cornu is fascinated by the processes of knowledge creation and the parallel between the epistemological obstacles in the past and the cognitive obstacles of students. Artigue continues the study to further levels of mathematical analysis and certain avenues of hope begin to appear. Tirosh reviews the cognitive conflicts inherent in contemplating the infinite and gives a detailed report of a single experiment exemplifying how students may be taught to reflect on their knowledge and actively participate in its reconstruction. Alibert and Thomas look at the process of proof and show the difficulties of the formalism and how it might be tackled through debate. Finally Dubinsky and I look forward to the use of the computer and the way in which this may change the nature of mathematics and provide an environment for learning. Despite the different tone of some chapters, the message of hope for reflective reconstruction of knowledge is there in all of them.

My recent thinking has led me to realize that the computer can be used in a very special way in learning – to carry out the processes, so that the user can concentrate on the product. This is the essence of a spread-sheet, a graph-plotter, a symbol manipulator, and so on. In other words, the computer allows a change in the encapsulation from process to object. Instead, of forcing the student first to learn and interiorize the process, the computer can carry out the process and allow the user to focus on the properties of the product. In this way there can be a shift of attention away from the process (in which the less able may become trapped) and towards the mathematical objects, and their relationships at a higher level. Instead of just learning the processes of *solving* differential equations, students may first appreciate the *existence* and *uniqueness* of solutions, and construct them in a meaningful, quasi-physical way, building an approximate solution curve by putting together short straight-line segments of the appropriate gradient.

Thus the final plank in the new charter of advanced mathematical thinking in the information age is what I have termed the *principle of selective construction of knowledge*, in which the learner is allowed, even encouraged, to focus separately on the processes of mathematics and the procepts produced by those processes. It is now possible to get a computer to carry out the algorithms so that the student can concentrate on the properties of the product. In this way the student can be encouraged to construct the properties and relationships enjoyed by the product whilst suppressing consideration of the process which is constructed internally by the computer. The student may at one time selectively concentrate purely on the process and at another on the higher level relationships. *Both* activities remain essential, for the process is needed to be able to *do* mathematics and the higher level relationships are essential to fit it together in a meaningful way. The interesting factor is that the focus on the process need not always precede the construction of the properties of the product. The intuitive idea of existence and uniqueness of differential equations can be investigated before formulating any symbolic solution. In this way the use of the computer gives new teaching and learning strategies in advanced mathematics.

We therefore arrive at a possible new synthesis in teaching and learning advanced mathematics which offers a more complete cycle of advanced mathematical thinking to students, even those of more modest abilities. The active participation in thinking is essential for the personal construction of meaningful concepts. Students need to be challenged to face the cognitive reconstruction explicitly, through conjecture and debate, through problem-solving, and they may be assisted in the acquisition of insights at higher levels by selectively sharing the construction with the computer. This does not remove the need to pass on information in the theorem-proof-application mode, for this is the crowning glory of advanced mathematics. But students need to be assisted through a transition to a stage where they see the necessity and economy of such an approach. Therefore, step by step, through professors being given a little space to experiment, initially as part of a traditional curriculum, a new balance may be struck, between the shining edifice of advanced mathematics that is the rightful pride of the mathematical community and the fuller range of advanced mathematical thinking that gave rise to its construction.

BIBLIOGRAPHY

Adler, C., (1966), 'The Cambridge conference report: blueprint or fantasy?', *Mathematics Teacher*, **58**, 210–17.

Alibert, D., (1987), 'Situation Codidactique et Délocalisation du Savoir', *Exposé au Séminaire de Didactique des Mathématiques et de l'Informatique*, Grenoble.

Alibert, D., (1988a), 'Towards New Customs in the Classroom', *For the Learning of Mathematics*, **8** (2), 31–35.

Alibert, D., (1988b), 'Codidactic System in the Course of Mathematics : How to Introduce it?', *Proceedings of the PME 12*, Veszprem.

Alibert, D., (1988c), 'How to Set Problems Before Giving Answers: An Experimental Mathematics Course at the University of Grenoble', *Proceedings of the Sixth International Congress of Mathematics Education, Budapest*.

Alibert, D., (1991), 'Sur le Rôle du Groupe–classe pour obtenir, et résoudre une situation a–didactique', *Recherches en Didactique des Mathématiques*, **10** (2), in press.

Alibert, D., Grenier, D., Legrand, M., Richard, F., (1986), 'Introduction du Débat Scientifique dans un Cours de Première Année du Deug A à l'Université de Grenoble 1', *Rapport de l'ATP 'Transitions dans le systeme educatif'* No. 122601 du MEN.

Alibert D., Legrand, M. & Richard, F., (1987a), 'Le thème "différentielles", un exemple de coordination maths–physique dans la recherche', *Actes du Colloque du Sèvres, Mai 1987*, Editions La Pensée Sauvage, Grenoble, 7–45.

Alibert, D., Legrand, M. & Richard, F., (1987b), 'Alteration of didactic contract in codidactic situations', *Proceedings of PME 11*, Montréal, 379–386.

Arton, H., (1981), *Elementary linear algebra*, John Wiley & Sons, New York.

Appel, K. & Haken, W., (1976), 'The solution of the four colour map problem', *Scientific American* (October), 108–121.

Arcavi, A. & Schoenfeld, A., (1987), *On the meaning of variable*, Technical Report, School of Education, University of California, Berkeley.

Artigue, M., (1987), 'Ingénierie didactique à propos d'équations differentielles', *Proceedings of PME 11*, Montréal, 236–242.

Artigue, M., Gautheron, V. & Isambard, E., (1985), 'Analyse non-standard et enseignement', *Cahier de Didactique No.15*, IREM Paris VII.

Artigue, M. & Szwed, T., (1983), *Représentations graphiques*, IREM Paris Sud.

Artigue, M. & Viennot, L., (1987), 'Students' conceptions and difficulties about differentials', *Proceedings of the Second International Seminar on Misconceptions and Educational Strategies in Science and Mathematics*, Cornell, III 1–7.

Atiyah, M.F., (1984), 'Mathematics and the computer revolution', *Nuovo Civilità della Macchine (Bologna)*, **2** (3), reprinted in A. G. Howson & J.-P. Kahane (Eds.), *The Influence of Computers and Informatics on Mathematics and its Teaching*, Cambridge University Press, Cambridge.

Ausubel, D.P., Novak, J. D. & Hanesian, H., (1968), *Educational Psychology, a Cognitive View*, (2nd edition), Holt, Rinehart & Winston, New York.

Authier, H., (1986), 'Étude comparative de diverses productions d'etudiants de première année de DEUG scientifique selon les séries de baccalauréat d'origine', *Cahier de Didactique No. 31*, IREM Paris VII.

Ayers, T., Davis, G., Dubinsky, E. & Lewin, P., (1988), 'Computer experiences in learning composition of functions', *Journal for Research in Mathematics Education*, **19** (3), 243–259.

Bachelard, G., (1938), (reprinted 1983), *La formation de l'esprit scientifique*, J. Vrin, Paris.

Balacheff, N., (1982), 'Preuve et Demonstration', *Recherche en Didactique des Mathematiques*, **3** (3), 261–304.

Balacheff, N., (1988), *Une Étude des Processus de Preuve en Mathématique chez des Élèves de Collége*, Thèse Université Grenoble 1.

Balacheff, N. & Laborde, C., (1985), 'Social Interactions for Experimental Studies of Pupils' Conceptions : Its Relevance for Research in the Didactics of Mathematics', *First International Conference on the Theory of Mathematics Education*, Bielefeld.

Bautier, E. & Robert, A., (1987), 'Apprendre des mathématiques et comment apprendre des mathématique', *Cahier de didactique des mathématiques 41*, IREM, Paris VII.

Beberman, M., (1956), 'The university of Illinois school mathematics program', in J. Bidwell & R. Clason (Eds.), *Readings in the History of Mathematics Education*, NCTM, Washington D.C., 655–663.

Begle, E., (1968), 'SMSG: The first decade', *Mathematics Teacher*, **62**, 239–45.

Beke, E., (1914), 'Rapport général sur les résultats obtenus dans l' introduction du calcul différentiel et intégral dans les classes supérieures des établissements secondaires', *L'Enseignement Mathématique*, **16**, 246–284.

Ben-Chaim, D., (1982), *Spatial visualization: sex differences, grade level differences and the effect of instruction on performance and attitudes*, Ph.D. dissertation, Michigan State University.

Berkeley, G., (1951), 'The Analyst', *Collected Works*, vol. 4 (ed. Luce, A. A. & Jessop, T.E.), Nelson, London.

Beth, E.W. & Piaget, J., (1966), *Mathematical Epistemology and Psychology* (W. Mays, trans.), Reidel, Dordrecht (originally published 1965).

Biggs, J. & Collis, K., (1982), *Evaluating the Quality of Learning: the SOLO Taxonomy*, Academic Press, New York.

Birkhoff, G.D., (1956), 'Mathematics of aesthetics', in J. R. Newman (Ed.), *The World of Mathematics Vol. 4*, (7th edition), Simon & Schuster, New York, 2185–2197.

Bishop, A., (1985), 'The Social Construction of Meaning – a Significant Development for Mathematics Education?', *For the Learning of Mathematics*, **5** (1), 24–28.

Bishop, E., (1977), review of *Elementary Calculus* by H. J. Keisler, *Bulletin American Mathematical Society*, **83**, 205–208.

Bishop, E., (1967), *Foundations of Constructive Analysis*, McGraw-Hill, New York.

Blackett, N., (1987), *Computer graphics and children's understanding of linear and locally linear graphs*, (M.Sc. Thesis), University of Warwick, U.K.

Boas, R., (1960), *A Primer of Real Functions*, Carus Mathematical Monographs, John Wiley and Sons, New York.

Bolzano, B., (1950), *Paradoxes of the Infinite*, Routledge & Kegan Paul, London.

Borasi, R., (1984), 'Some reflections on and criticisms of the principle of learning concepts by abstraction', *For the Learning of Mathematics*, **4**, 14–18.

Borasi, R., (1985), 'Errors in the enumeration of infinite sets', *Focus on Learning Problems in Mathematics*, **7**, 77–89.

Boyer, C. B., (1939), *The History of the Calculus and its Conceptual Development*, (page references as in reprint, Dover, New York, 1959).

Breuer, S., Gal-Ezer, J. & Zwas, G., (1990), 'Microcomputer Laboratories in Mathematics Education', *Computers in Mathematics and its Applications*, **19** (3), 13–34.

Brousseau, G., (1986), 'Fondements et Méthodes de la Didactique des Mathématiques', *Recherches en Didactique des Mathématiques*, **7** (2), 33–115.

Brousseau, G., (1988), *Fondements et méthodes de la didactique des mathématiques*, Thesis, University of Bordeaux.

Brown, A. L., Bransford, J. D., Ferrara, R. A. & Campione, J. C., (1983), 'Learning, Remembering and Understanding', in P. H. Mussen (Ed.), *Handbook of Child Psychology*, Fourth Edition, Vol. III: *Cognitive Development* (volume eds. J. H. Flavell & E. M. Markman), Wiley, New York, 77–166.

Brown, A. L. & Kane, M. J., (1988), 'Pre-school children can learn to transfer: Learning to learn and learning from example', *Cognitive Psychology*, 20, 493–523.

Buck, R. C., (1970), '"Functions" in mathematics education', in E. G. Begle (Ed.), *69th Yearbook of the National Society for the Study of Education*, University of Chicago Press, Chicago, 236–259.

Cambridge Conference, (1963), *Goals for School Mathematics*, (Report of the Cambridge Conference on School Mathematics) Houghton Mifflin Co., Boston, for Educational Services, Inc., 10.

Cambridge Conference, (1967), *Goals for Mathematics Education for Elementary School Teachers*, (Cambridge Conference on Teacher Training), Houghton Mifflin Co., Boston, for Educational Development Center, Inc., 98.

Cajori, F., (1929), *A history of mathematical notations*, Vol. 2: *Notations mainly in higher mathematics*, The Open Court Publishing Co, La Salle, Illinois.

Cajori, F., (1980), *History of Mathematics*, (Third edition, originally published 1893), Chelsea Publishing Co., New York.

Cantor, G., (1955), *Contributions to the Founding of the Theory of Transfinite Numbers*, Dover, New York.

Carter, H. C., (1970), 'A study of one learner cognitive style and the ability to generalize behavioral competences', *Proceedings of AERA*, ERIC No. ED 040–758.

Case, R. & Sandleson, R., (1988), 'A developmental approach to the identification and teaching of central conceptual structures in middle school science and mathematics', in J. Hiebert & M. Behr (Eds.), *Number Concepts and Operations in the Middle Grades*, Erlbaum, Hillsdale NJ.

Char, B. W., Fee, G. J., Geddes, K. O., Gonnet, G. H., Marshman, B. J. & Ponzo, P., (1986), 'Computer Algebra in the Mathematics Classroom', *Proc. 1986 Symposium on Symbolic and Algebraic Computation, Assoc. Comput. Mach.*, 135–140.

Cheshire, F.D., (1981), *The effect of learning computer programming skills on developing cognitive abilities*, Doctoral dissertation, Arizona State University, *Dissertation Abstracts International*, 42, 654A, (University Microfilms No.81–17163).

Cipra, B., (1983), *Misteaks: (Mistakes) A Calculus Supplement*, Birkhauser, Boston.

Clement, J., Lochhead, J. & Monk, J. S., (1981), 'Translation difficulties in learning mathematics', *American Mathematical Monthly*, 286–290.

Clement, J., (1983), 'A conceptual model discussed by Galileo and used intuitively by physics students', in D. Gentner and A. L. Stevens (Eds.), *Mental Models*, Lawrence Erlbaum Associates, London, 325–340.

Clement, J., (1986), 'Misconceptions in graphing', *Proceedings of PME 9*, Noordwijkerhout, 369–375.

Clement, J., (1987), 'Overcoming students' misconceptions in physics: The role of anchoring intuitions and analogical validity', in J. D. Novak (Ed.), *Proceedings of the Second International Seminar on Misconceptions and Educational Strategies Strategies in Science and Mathematics*, 3, 84–97. Cornell University, Ithaca NY.

Clements, M. A., (1984), 'Terence Tao', *Educational Studies in Mathematics*, 15 (21), 32–38.

Cohen, L. & Ehrlich, G., (1963), *The Structure of the Real Number System*, D. van Nostrand Co. Inc, Princeton.

Cornu, B., (1981), 'Apprentissage de la notion de limite: modèles spontanés et modèles propres', *Actes du Cinquième Colloque du Groupe Internationale PME*, Grenoble, 322–326.

Cornu, B., (1983), *Apprentissage de la notion de limite: conceptions et obstacles*, Thèse de doctorat de troisième cycle, L'Université Scientifique et Medicale de Grenoble.

D' Alembert, J. L., (1975), 'Differentiel', *Encyclopedie Méthodique Mathematiques*, 1, 520–560, Panc Koucke Ed., Paris.

D' Halluin, C. & Poisson, D., (1988), *Une stratégie d'enseignement des mathématiques: la mathémematisation de situations intégrant l'informatique comme outil et mode de pensée*, Thèse de Doctorat, Université de Lille.

Dauben, J., (1983), 'George Cantor and the origins of transfinite set theory', *Scientific American*, 248, 112–121.

Davis, P. J., (1986), 'The nature of proof', in M.Carss (Ed.) *Proceedings of the fifth international congress on mathematical education*, Birkhauser, Boston.

Davis, P. J., (1972), 'Fidelity in mathematical discourse: is one and one really two?', *American Mathematical Monthly*, 79, 252–263.

Davis, P. J. & Hersh, R., (1981), *The Mathematical Experience*, Birkhauser, Boston.

Davis, P. J. & Hersh, R., (1986), *Descartes' Dream. The World According to Mathematics*, Houghton Mifflin Co, Boston.

Davis, R. B., (1984), *Learning Mathematics: The Cognitive Science Approach to Mathematics*, Ablex, Norwood NJ.

Davis, R. B., (1986), 'Algebra in elementary schools', in M. Carss (Ed.), *Proceedings of the Fifth International Congress on Mathematical Education*, Birkhauser, Boston.

Davis, R. B., (1988), 'The Interplay of Algebra, Geometry, and Logic', *Journal of Mathematical Behavior*, 7, 9–28.

Davis, R. B. & Vinner, S., (1986), 'The Notion of Limit: Some Seemingly Unavoidable Misconception Stages', *Journal of Mathematical Behaviour*, 5 (3), 281–303.

Delens, P., (1930), 'Notations différentielles', *L'Enseignement Mathématique*, 30, 333–337.

Deligne, P., (1977), 'Seminaire de Géométrie Algébrique du Bois Marie (SGA4) Théorie des Topos et Cohomologie Étale des Schémas', Lecture Notes in Mathematics no. 305, vol. 3, Springer-Verlag, Berlin.

Dienes, Z. P., (1960), *Building up Mathematics*, Hutchinson, London.

Dienes, Z. P., (1968), *Fractions: an operations approach*, Hutchinson, London.

Dienes, Z. P. & Jeeves, J., (1965), *Thinking in Structures*, Hutchinson, London.

Dieudonne, J. A., (1971), 'Modern axiomatic methods and the foundations of mathematics', in F. Le Lionnais (Ed.), *Great currents of mathematical thought* (Vol. 1), Dover, New York.

Dieudonné, J. A., (1975), 'L'abstraction et l'intuition mathématique', *Actes du Colloque sur Les Mathématiques et la Réalité, Centre Universitaire de Luxembourg, 1974*, published in *Dialectica, Revue internationale de philosophie de la connaissance*, 29 (1), 39–54.

Donaldson, M., (1963), *A study of children's thinking*, Tavistock Publications, London, 183–185.

Dörfler, W., (1988), 'Die Genese mathematischer Objekte und Operationen aus Handlungen als kognitive Konstruktion', in Willibald Dörfler (Ed.), *Kognitive Aspekte mathematischer Begriffsentwicklung*, Hölder-Pichler-Tempsky, Vienna, Austria, 55–125.

Douady, R., (1984), 'L'ingénierie didactique, une instrument privilegie pour une prise en compte de la complexité de la class', *Proceedings of PME 11*, Montréal, III 222–228.

Douady, R., (1986), 'Jeu de Cadres et Dialectique Outil-objet', *Recherches en Didactique des Mathématiques*, 7 (2), 5–32.

Dreyfus T., (1991), 'On the status of visualization and visual reasoning in mathematics and mathematics education', *Proceedings of PME 15*, Assisi, 1, 33–48.

Dreyfus, T. & Vinner, S., (1982), 'Some aspects of the function concept in college students and junior high school teachers', *Proceedings of PME 6*, Antwerp, 12–17.

Dreyfus, T. & Eisenberg, T., (1983), 'The function concept in College Students: Linearity, Smoothness and Periodicity', *Focus on Learning Problems in Mathematics*, 5 (3 & 4), 119–132.

Dreyfus, T. & Eisenberg, T., (1984), 'Intuitions on functions', *Journal of Experimental Education*, 52, 77–85.

Dreyfus, T. & Eisenberg, T., (1986), 'On the Aesthetics of Mathematical Thought', *For the Learning of Mathematics*, 6 (1), 2–10.

Dreyfus, T. & Eisenberg, T., (1987), 'On the deep structure of functions', *Proceedings of PME 11*, Montréal, 190–96.

Dreyfus, T. & Thompson, P. W., (1985), 'Microworlds and Van Hiele Levels', *Proceedings of PME 9*, Noordwijkerhout, 1, 5–11.

Dubinsky, E., (1986), 'Teaching mathematical induction I', *The Journal of Mathematical Behavior*, 5, 305–317.

Dubinsky, E., (1989), 'The case against visualization in school and university mathematics', position paper presented to the Advanced Mathematical Thinking Group at PME 13, Paris, (Available from the author: Department of Mathematics, Purdue University, West Lafayette, Indiana, USA).

Dubinsky, E., (1990a), 'On learning quantification', *Journal of Mathematical Behavior*, (in press).

Dubinsky, E., (1990b), 'Teaching mathematical induction II', *The Journal of Mathematical Behavior*, (in press), 285–304.

Dubinsky, E. & Lewin, P., (1986), 'Reflective abstraction and mathematics education: the genetic decomposition of induction and compactness', *The Journal of Mathematical Behavior*, 5, 55–92.

Dubinsky, E. & Schwingendorf, K. E., (1990a), 'Calculus, concepts and computers – innovations in learning calculus', *CRAFTY* (ed. Tucker T.), Math. Assoc. Amer.

Dubinsky, E. & Schwingendorf, K. E., (1990b), 'Constructing calculus concepts: cooperation in a computer laboratory', *MAA Notes Series*, (ed. Leinbach C.), Math. Assoc. Amer.

Dubinsky E., Elterman, F. & Gong, C., (1988), 'The student's construction of quantification', *For the Learning of Mathematics* , 8 (2), 44–51..

Dubinsky E., Hawks, J., & Nichols, D., (1989), 'Development of the Process Conception of Function in Pre–Service Teachers in a Discrete Mathematics Course', *Proceedings of the PME 13*, Paris.

Dubinsky, E, & Lewin, P., (1986), 'Reflective Abstraction and Mathematics Education: the genetic decomposition of induction and compactness', *The Journal of Mathematical Behavior*, 5, 55–92.

Duval, R., (1983), 'L'Obstacle du dedoublement des objets mathématiques', *Educational Studies in Mathematics*, 14, 385–414.

Edwards, E. M., (1987), 'An appreciation of Kronecker', *The Mathematical Intelligencer*, 9 (1), 28–35.

Ellerton, N. F., (1985), *The Development of Abstract Reasoning – Results from a large scale mathematics study in Australia and New Zealand.*

Fehr, H., (1966), A unified mathematics program for grades seven though twelve, *The Mathematics Teacher*, 59, 463.

Fehr, H., (1974), 'The secondary school mathematics curriculum improvement study: a unified mathematics program', *Mathematics Teacher*, 25–33.

Fischbein, E., (1982), 'Intuition and Proof', *For the Learning of Mathematics*, 3 (2), 9–18, 24.

Fischbein, E., (1978), 'Intuition and mathematical education', *Osnabrücker Schriften zür Mathematik* 1, 148–176.

Fischbein, E. & Gazit, A., (1984), 'Does the teaching of probability improve probabilistic intuitions?', *Educational Studies in Mathematics*, 15, 1–24.

Fischbein, E., Tirosh, D. & Hess, P., (1979), 'The Intuition of Infinity', *Educational Studies in Mathematics*, 10, 3–40.

Fischbein, E., Tirosh, D. & Melamed, U., (1982), 'Is it possible to measure the intuitive acceptance of the mathematical statement?', *Educational Studies of Mathematics*, 12, 491–512.

Fodor, J. A., Garret, M. F., Walker, E. C. & Parley, C. H., (1980), 'Against definition', *Cognition*, **8**, 263–267.

Frankel, A. A., (1953), *Abstract Set Theory*, North-Holland Pub., Amsterdam.

Fréchet, M., (1911), 'Sur la notion de différentielle', *Comptes-Rendus de l'Académie des Sciences*, **152** (13), 845–847, 1050–1051.

Freudenthal, H., (1983), *The Didactical Phenomenology of Mathematics Structures*, Reidel, Dordrecht.

Gagné, R. M., (1970), *The Conditions of Learning*, 2nd ed., Holt, Rinehart and Winston, New York.

Galileo, G., (1954), *The New Sciences*, (H. Crew and A. Salvio, Trans.), Dover, New York (previously published 1811).

Gazzanigna, M. S., (1985), *The Social Brain: Discovering Networks of the Mind*, Basic Books, New York.

Gleick, J., (1987), *Chaos: Making a New Science*, Penguin, London.

Glennon, V. J., (1980), 'Neuropsychology and the Instructional Psychology of Mathematics', *The Seventh Annual Conference of the Research Council for Diagnostic and Prescriptive Mathematics*, Vancouver, B.C.

Goldin, G., (1982), 'Mathematical language and problem solving', *Visible language* (Special issue on mathematical language, R. Skemp, Ed.), **16**, 221–238.

Gray, E. M. & Tall D. O., (1991), 'Duality, Ambiguity and Flexibility in Successful Mathematical Thinking', *Proceedings of PME 15*, Assisi, **2**, 72–79.

Gray E. M., (1991), 'An Analysis of Diverging Approaches to Simple Arithmetic : Preference and its Consequences', *Educational Studies in Mathematics*, (in press).

Greeno, G. J., (1983), 'Conceptual entities', in D. Genter & A. L. Stevens (Eds.), *Mental Models*, 227–252.

Grenier, D., Legrand, M. & Richard, F., (1984), 'L'Introduction du Débat Scientifique à l'Intérieur du Cours pour Provoquer chez les Étudiants un Processus du Découverte et de Preuve', Paper presented at the *Séminaire de Didactique*, Grenoble.

Grenier, D., Legrand, M. & Richard, F., (1985), 'Une Séquence d'Enseignement sur l'Intégrale en DEUG A Première Année', *Cahier de Didactique des Mathématiques No. 22*, IREM Paris VII.

Grogono, P., (1989), 'Meaning and Process of Mathematics and Programming', *For the Learning of Mathematics*, **9**, 14–19.

Hadamard, J., (1945), *The Psychology of Invention in the Mathematical Field.*, Princeton University Press, (page references are to the Dover edition, New York 1954).

Hadar-Moscovitz, N., Zaslavsky, O., & Inbar, S., (1987), 'An empirical classification model for errors in high school mathematics', *Journal for Research in Mathematics Education*, 3–14.

Hahn, H., (1956), 'Infinity', in J. R. Newman (Ed.), *The World of Mathematics*, Vol. 3, Simon & Schuster, New York, 1593–1611.

Halmos, P. R., (1985), *I want to be a mathematician*, Springer-Verlag, New York.

Hammersley, J. M., (1968), 'On the enfeeblement of mathematical skills by "modern mathematics" and by similar soft intellectual trash', *Bulletin of the Institute of Mathematics and Its Applications*, (Oct., 1968), 66–85.

Hanna, G., (1983), *Rigorous proof in mathematics education*, OISE Press, Toronto.

Harel, G., (1985), *Teaching linear algebra in high-school*, unpublished doctoral dissertation, Ben-Gurion University of the Negev, Beer-Sheva, Israel.

Harel, G., (1987), 'Variation in linear algebra content presentations', *For the Learning of Mathematics*, **7**, 29–31.

Harel, G., (1989a), 'Learning and teaching linear algebra: Difficulties and an alternative approach to visualizing concepts and processes', *Focus on Learning Problems in Mathematics*, **11**, 139–148.

Harel, G., (1989b), 'Applying the principle of multiple embodiments in teaching linear algebra: aspects of familiarity and mode of representation', *School Science and Mathematics*, (in press).

Harel, G. & Tall, D. O., (to appear): 'The General, The Abstract and the Generic in Advanced Mathematical Thinking', *For the Learning of Mathematics*.

Hart, K. M. (ed), (1981), *Children's Understanding of Mathematics 11–16*, John Murray, London.

Heid, K., (1984), *Resequencing Skills and Concepts in Applied Calculus through the Use of the Computer as a Tool*, Ph.D. Thesis, Pennsylvania State University.

Henle, J. M. & Kleinberg, E. M., (1979), *Infinitesimal Calculus*, MIT Press, Cambridge.

Hiebert, J., (1986), *Conceptual and Procedural Knowledge*, Erlbaum, Hillsdale NJ.

Hilbert, D., (1981), 'Axiomatisches Denken', *Mathematische Annalen*, **78**, 405–415.

Hilbert, D., (1964),' On the infinite' in P. Benacerraf & H. Putman (Eds.), *Philosophy of Mathematics* (34–151), Prentice-Hall, New Jersey (original work published 1923).

Hodgson, B. R., (1987), 'Symbolic and Numerical Computation: the computer as a tool in mathematics', in D. C. Johnson & F. Lovis (Eds.), *Informatics and the Teaching of Mathematics*, North-Holland, Amsterdam, 55–60.

Hoffman, K. M., (1989), *The Science of Patterns: A Practical Philosophy of Mathematics Education*, Lecture to SIG/RME, AERA Annual Meeting, San Francisco, CA.

Hubbard, J. H. & West, B. H., (1985), 'Computer Graphics Revolutionize the Teaching of Differential Equations', *Supporting Papers for the ICMI Symposium*, IREM, Université Louis Pasteur, Strasbourg.

Hubbard, J. H. & West, B. H., (1989), *Ordinary differential equations*, Springer-Verlag, New York.

Janvier, C., (1978), *The interpretation of complex cartesian graphs representing situations – studies and teaching experiments*, (Doctoral dissertation), University of Nottingham, England.

Kaput, J. J., (1982), 'Differential effects of the symbol systems of arithmetic and geometry on the interpretation of algebraic symbols', paper presented at the meeting of the American Educational Research Association, New York.

Kaput, J. J., (1987a), 'Towards a theory of symbol use in mathematics', in C. Janvier (Ed.) *Problems of representation in mathematics learning and problem solving*, Erlbaum, Hillsdale NJ.

Kaput, J. J., (1987b), 'PME 11 Algebra papers: A Representational Framework', *Proceedings of PME11*, Montréal, **1**, 345–354.

Kaput, J. J., (1989), 'Linking representations in the symbol system of algebra', in C. Kieran & S. Wagner (Eds.) *A research agenda for the teaching and learning of algebra*, NCTM, Reston, VA and Erlbaum, Hillsdale NJ.

Kaput, J. J. (in press), 'Notations and representations as mediators of constructive processes', in E. von Glasersfeld (Ed.), *Constructivism in mathematics education*, Reidel, Dordrecht.

Kaput, J. J. (in press), 'Creating cybernetic and psychological ramps from the concrete to the abstract: Examples from multiplicative structures', to appear in J. Schwartz, M. West., M. S. Wiske & D. Perkins (Eds.) *Making sense of the future: Technology in mathematics and science education*, Harvard University Press, Cambridge, MA.

Kaput, J. J. (in preparation), 'Patterns in students: Formalization of quantitative patterns', to appear in G. Harel & E. Dubinsky (Eds.), *The Concept of Function: Aspects of epistemology and pedagogy*, MAA.

Kaput, J. J., West, M. M., Luke, C. & Pattison-Gordon, L., (1988), 'Concrete representations for ratio reasoning', *Proceedings of the Tenth Annual Meeting of the North American Chapter of the International Group for the Psychology of Mathematics Education*, DeKalb, IL: Northern Illinois University, 93–99.

Karplus, R., (1979), 'Continuous functions: students' viewpoints', *European Journal of Science Education*, **1**, 397–415.

Kautschitsch, H., (1988), 'Bild-unterstützte Abstraktion und Verallgemeinerung', in W. Dörfler (Ed.), *Kognitive Aspekte mathematischer Begriffsentwicklung*, Hölder-Pichler-Tempsky, Vienna,

Austria, 191–258.

Keisler, H. J., (1971), *Elementary calculus: an approach using infinitesimals*, Prindle, Weber & Schmidt, Boston.

Keisler, H. J., (1976), *Elementary Calculus*, Prindle, Weber & Schmidt, Boston.

Kitcher, P., (1984), *The nature of mathematical knowledge*, Oxford University Press, New York.

Kline W., Oesterle R., & Willson, L., (1959), *Foundations of Advanced Mathematics*, American Book Co., New York, 239.

Kline, M., (1958), 'The ancients versus the moderns: A new battle of the books', *Mathematics Teacher*, **51**, 418–27.

Kline, M., (1970), 'Logic versus pedagogy', *American Mathematics Monthly*, **77**, 264–282.

Kleiner, I., (1989), 'Evolution of the function concept: A brief survey', *The College Mathematics Journal*, **20** (4), 282–300.

Koçak, H., (1986), *Differential and difference equations through computer experiments*, Springer-Verlag, New York.

Kolata, G., (1984), 'Surprise Proof of an Old Conjecture', *Science*, **225**, 1006–1007.

Krutetskii, V. A., (1976), *The Psychology of Mathematical Abilities in School Children*, J. Teller (*transl*), J. Kilpatrick & I. Wirszup (Eds.), University of Chicago Press, Chicago.

L'Hospital, Marquis de, (1696), *Analyse des infinient petits pour l'intelligence des lignes courbes*, Paris.

Lakatos, I. M., (1976), *Proofs and refutations: The logic of mathematical discovery*, Cambridge University Press, Cambridge.

Lakatos, I. M., (1978), 'Cauchy and the Continuum: The Significance of Non-Standard Analysis for the History and Philosophy of Mathematics', *Mathematical Intelligencer*, **1** (3), 151–161.

Lander, L. J. & Parkin, T. R., (1967), 'A counter-example to Euler's sum of powers conjecture', *Math. Comp.*, **21**, 101–103.

Lane, K. D., Ollongren, A. & Stoutmyer, D., (1986), 'Computer Based Symbolic Mathematics for Discovery', *The Influence of Computers and Informatics on Mathematics and its Teaching*, (ed. Howson A. G. & Kahane J.-P.), Cambridge University Press, Cambridge, 133–146.

Laurent, H., (1899), 'Considérations sur l'enseignement des mathématiques dans les classes de Spéciales en France', *L'Enseignement Mathématique*, **1**, 38–44.

Legrand, M., *et al*, (1988), 'Le débat scientifique', *Actes du colloques franco-allemands de Marseille*, 53–66.

Legrand, M., *et al*, (1986), *Introduction du débat scientifique dans un cours de première année de DEUG A à l'Université de Grenoble I*, Rapport de recherche, Editions IMAG, Grenoble.

Legrand, M. & Richard, F., (1984), 'Mathématiques Expérimentales ou une Approche de la Conjecture et de la Preuve par l'Étudiant', *Colloque "Ingenierie Pédagogique dans l'Enseignement Superieur"*, Paris.

Lehman, D. R., Lempert, R. O. & Nisbett, R. E., (1988), 'The effects of graduate training on reasoning: formal discipline and thinking about everyday-life events', *American Psychologist*, **43**, 431–442.

Leinhardt, G., Zaslavsky, O. & Stein, M., (1990), '"Functions, Graphs, and Graphing: Tasks, Learning, and Teaching", review of Educational Research', **60** (1), 1–64.

Leron, U., (1983a), Structuring Mathematical Proofs, *The American Mathematical Monthly*, **90** (3), 174–184.

Leron, U., (1983b), 'The Diagonal Method', *Mathematics Teacher*, **76** (9), 674–676.

Leron, U., (1985a), 'Heuristic Presentations: the Role of Structuring', *For the Learning of Mathematics*, **5** (3), 7–13.

Leron, U., (1985b), 'A Direct Approach to Indirect Proof', *Educational Studies in Mathematics*, **16**, 321–325.

Lorenz, E., (1963), 'Deterministic Nonperiodic Flow', *Journal of the Atmospheric Sciences*, **20**, 448–464.

MacLane, S., (1971), *Categories for the working mathematician*, Springer-Verlag, New York.

MacLane, S., (1986), 'Criteria for excellence in mathematics', *Bull. de la Soc. Math. de Belg. Série A*, **38**, 301–302.

MacLane, S., (1965), *Proceedings of the Preliminary Meeting on College Level Mathematics* (Katada Report 1964), Japanese International Printer.

Major, R. & Clark, C., (1963), 'Explorations in students controlled instruction', in G. Fiesh & H. Meier (Eds), *National Society for Programmed Instruction*.

Manin, Y. I., (1977), *A course in mathematical logic*, Springer-Verlag, New York.

Martin, G. & Wheeler, M. M., (1987), 'Infinity concepts among pre-service elementary school teachers', *Proceedings of PME11*, Montréal, **3**, 362–368.

Maslow, A. H., (1970), *Motivation and personality*, Harper & Row, New York.

Mason, J. with Burton, L. & Stacey, K., (1982), *Thinking Mathematically*, Addison-Wesley, London.

Mason, J., (1989), 'Mathematical Abstraction as the result of a delicate shift of attention', *For the Learning of Mathematics*, **9** (2), 2–8.

McCloskey, M., (1983), 'Intuitive physics', *Scientific American*, **248**, 114–122.

Menis, Y., Snyder, M. & Ben-Kohav, E., (1980), 'Improving achievement in algebra by means of the computer', *Educational Technology*, **20**, 19–22.

Monaghan, J. D., (1986), *Adolescents' Understanding of Limits and Infinity*, unpublished Ph.D. thesis, Warwick University.

Movshovitz-Hadar, N., (1988), 'Stimulating Presentation of Theorems Followed by Responsive Proofs', *For the Learning of Mathematics*, **8** (2), 12–19.

Muir, A., (1988), 'The Psychology of Mathematical Creativity', *Mathematical Intelligencer*, **10** (1), 33–37.

Mundy, J., (1984), 'Analysis of errors of first year calculus students', in *Theory, Research and Practice in Mathematics Education*, A. Bell, B. Low & J. Kilpatrick (Eds.), *Proceedings of ICME 5*, Adelaide, Working group reports and collected papers, Shell Centre, Nottingham, U.K., 170–172.

Munroe, M. E., (1965), *Introductory Real Analysis*, Addison-Wesley, Atlanta.

Nelson, E., (1977), 'Internal set theory – a new approach to non-standard analysis', *Bulletin Americal Mathematical Society*, **83**, 1165–1198.

Olson, D. R. & Campbell, R., (in press), 'Representation and Misrepresentation: On the beginning of symbolization in young children', in D. Tirosh (Ed.), *Implicit and Explicit Knowledge: An Educational Approach*, Ablex, Norwood NJ.

Orton, A., (1980), *A cross-sectional study of the understanding of elementary calculus in adolescents and young adults*, unpublished Ph.D. Thesis, Leeds University, U.K.

Orton, A., (1983a), 'Students' understanding of integration', *Educational Studies in Mathematics*, **14** (1), 1–18.

Orton, A.,, (1983b), 'Students' understanding of differentiation', *Educational Studies in Mathematics*, **14** (3), 235–250.

Papert, S., (1980), *Mindstorms*, Basic Books, New York & Harvester Press, Brighton, U.K.

Paulos, J. A., (1988), *Innumeracy: Mathematical Illiteracy and its Consequences*, Hill & Wang, New York.

Peitgen, H. & Jürgens, H., (1989), 'Fraktale: Computerexperimente (ent)zaubern komplexe Strukturen', *Mathematik-Unterricht*, **5**, 4–19.

Phillips, E. G., (1931), 'The teaching of differential', *The Mathematical Gazette*, **15**, 401–403.

Piaget, J., (1952), *The Child's Conception of Number*, Norton, New York (original published 1941).

Piaget, J., (1970a), *Genetic Epistemology* (E. Duckworth, trans.), Columbia University Press, New York.

Piaget, J., (1970b), *Structuralism* (C. Maschler, trans.), Basic Books, New York (original published 1968).

Piaget, J., (1971), *Biology and Knowledge* (B. Walsh, trans.), University of Chicago Press, Chicago (original published 1967).

Piaget, J., (1972a), *The principles of Genetic Epistemology* (W. Mays trans.), Routledge & Kegan Paul, London (original published 1970).

Piaget, J., (1972b), 'Comments on Mathematical Education', in A.J. Howson (Ed.), *Developments in Mathematical Education, Proceedings of the Second International Congress in Education,* Cambridge University Press, Cambridge.

Piaget, J., (1975a), 'Piaget's Theory', in P. B. Neubauer (Ed.), *The Process of Child Development,* Jason Aronson, New York, 164–212.

Piaget, J., (1975b), 'Piaget's theory', in P. H. Mussen (Ed.), *Carmichael's Manual of Child Psychology* (Vol. 1), John Wiley & Sons, New York.

Piaget, J., (1976), *The Grasp of Consciousness* (S. Wedgwood, trans.), Harvard University Press, Cambridge MA (original published 1974).

Piaget, J., (1978), *Success and Understanding* (A. J. Pomerans, trans.), Harvard University Press, Cambridge MA (original published 1974).

Piaget, J., (1980), *Adaptation and Intelligence* (S. Eames, trans.), University of Chicago Press, Chicago (original published 1974).

Piaget, J., (1985), *The Equilibration of Cognitive Structures* (T. Brown and K. J. Thampy, trans.), Harvard University Press, Cambridge MA (original published 1975).

Piaget, J. & Garcia, R., (1983), *Psychogenèse et histoire des sciences,* Flammarion, Paris.

Piaget, J., Grize, J.-B., Szeminska, A., & Bang, V., (1977), *Epistemology and Psychology of Functions* (J. Castellanos & V. Anderson, trans.), Reidel, Dordrecht (original published 1968).

Piaget, J., Inhelder, B., & Szeminska, A., (1960), *The Child's Conception of Geometry* (E. A. Lunzer, trans.), Norton, New York.

Pimm, D., (1984), 'The role of the individual in mathematics texts', *Proceedings of PME 8,* Sydney, 469–77.

Poincaré, H., (1899), 'La notation différentielle et l'enseignement', *L'Enseignement Mathématique,* **1**, 106–110.

Poincaré, H., (1913), *The Foundations of Science* (translated by Halsted G.B.), The Science Press, New York (page references as in University Press of America edition, 1982).

Poincaré, H., (1963), *Mathematics and Science: Last Essays,* (J. Bolduc trans.), Dover, New York, (original published 1913).

Polya, G., (1945), *How to Solve It,* Princeton University Press, Princeton.

Polya, G., (1954), *Mathematics and Plausible Reasoning,* (2 volumes), Princeton University Press, Princeton.

Polya, G., (1966), 'On teaching problem solving', in *The Role of Axiomatics and Problem Solving in Mathematics,* Ginn & Co., Washington, D.C., 123–29.

Polya, G., (1980), *Mathematical Discovery,* (2 volumes), Academic Press, New York.

Ponte, J., (1984), *Functional reasoning and the interpretation of cartesian graphs,* (Doctoral dissertation), University of Georgia, Athens, Georgia.

Reding, A. H., (1981), *The effects of computer programming on problem solving abilities of fifth grade students,* Doctoral dissertation, University of Wyoming, Dissertation Abstracts International, 42, 3484A, (University Microfilms No. 82–01793).

Riley, M. S., Greeno, J. G. & Heller, J. L., (1983), 'Development of children's problem solving ability in arithmetic', in H. P. Ginsburg (Ed.), *The development of mathematical thinking,* Academic Press, New York, 153–196.

Rival, I., (1987), 'Picture puzzling: mathematicians are rediscovering the power of pictorial reasoning', *The Sciences*, 41–46.

Robert, A., (1982a), *L'Acquisition de la notion de convergence des suites numériques dans l'Enseignement Supérieur*, Thèse de Doctorat d'État, Paris VII.

Robert, A., (1982b), 'L'Acquisition de la notion de convergence des suites numériques dans l'Enseignement Supérieur', *Recherches en Didactique des Mathématiques*, 3 (3), 307–341.

Robert, A., (1984), 'Connaisances des élèves sur les debuts de l'analyse sur ℝ à la fin des études scientifiques secondaire françaises', *Cahier de didactique des mathématiques 18*, IREM, Paris VII.

Robert, A., (1985), 'Rapports enseignement/apprentissage (débuts de l'analyse sur ℝ) – Analyse d'une section de DEUG A première année, *Cahier de didactique des mathématiques 18-1*, IREM, Paris VII.

Robert, A. & Boschet, F., (1984), 'L'acquistion des débuts de l'analyse sur ℝ dans un section ordinaire de DEUG première année', *Cahier de didactique des mathématiques 7*, IREM, Paris VII.

Robert, A. & Tenaud, I., (1988), 'Une expérience d'enseignement de la géometrie en Terminal C', *Recherches en Didactique des Mathématiques*, 9 (1) 31–70.

Robert, A., Rogalski, J. & Samurçay, R., (1987), 'Enseigner des méthodes', *Cahier de didactique des mathématiques 38*, IREM, Paris VII.

Roberti, J. V., (1987), 'The indirect method', *Mathematics Teacher*, 80, 41–43.

Robinet, J., (1984), *Ingénierie didactique de l'elementaire au superieur*, University of Paris VII.

Robinson, A., (1966), *Non-Standard analysis*, North Holland, Amsterdam, Holland.

Robitaille, D. F, Sherrill, J. M. & Kaufman, D. M., (1977), 'The effect of computer utilization on the achievement and attitudes of grade nine mathematics students', *Journal for Research in Mathematics Education*, 8, 26–32.

Rucker, R., (1982), *Infinity and the mind*, Birkhauser Boston Inc., Cambridge, Ma.

Russell, B. R., (1956), 'Mathematics and the Metaphysicians', in J. R. Newman (Ed.), *The World of Mathematics* (Vol. 3, 1576–1592), Simon & Schuster, New York.

Ruthven, K., (1990), 'The influence of graphic calculator use on translation from graphic to symbolic forms', *Educational Studies in Mathematics*, 21, 5, 431–450.

Sawyer, W. W., (1987), 'Intuitive Understanding of Mathematical Proof', *Bulletin of the I.M.A.*, 23, 61–62.

Schoenfeld, A. H., (1983), 'Beyond the Purely Cognitive : Belief Systems, Social Cognitions and Metacognition as Driving Forces and Intellectual Performance', *Cognitive Science*, 7, 329–363.

Schoenfeld, A. H., (1985), *Mathematical Problem Solving*, Academic Press, Orlando.

Schoenfeld, A. H., (1986), 'On having and using geometric knowledge', in J. Hiebert (Ed.), *Conceptual and Procedural Learning*, Lawrence Erlbaum, Hillsdale NJ.

Schoenfeld, A. H., Smith, J. P. & Arcavi, A., (to appear), 'Learning: the microgenetic analysis of one students evolving understanding of a complex subject matter domain', in R. Glaser (ed.), *Advances in Instructional Psychology* (vol.4), Erlbaum, Hillsdale NJ.

Schoenfield, A. H., (1989), 'Explorations of Students' Mathematical Beliefs and Behavior', *Journal for Research in Mathematics Education*, 20 (4), 338–335.

Schoenfeld, A. H. (in press), 'On mathematics as sense-making: an informal attack on the unfortunate divorce of formal and informal mathematics', in D. Perkins, J. Segal & J. Voss (Eds.), *Informal Reasoning and Education*, Lawrence Erlbaum, Hillsdale NJ.

School Mathematics Project, (1967), *Advanced Mathematics Book I*, C.U.P., Cambridge.

Schwarz, B., (1989), *The use of a microworld to improve ninth grade' concept image of a function: The Triple Representation Model curriculum*, PhD thesis, Weizmann Institute of Science, Rehovot, Israel.

Schwarz, B., Dreyfus, T. & Bruckheimer, M., (1990), 'A model of the function concept in a three-fold representation', *Computers and Education*, **14** (3), 249–262.

Schwarz, B. & Dreyfus, T., (1991), 'Assessment of thought processes with mathematical software', *Proceedings of PME 15*, Assisi.

Schwarzenberger, R. L. E. & Tall, D. O., (1978), 'Conflicts in the learning of real numbers and limits', *Mathematics Teaching*, **82**, 44–49.

Selden, J., Mason, A. & Selden, A., (1989), 'Can Average Calculus Students Solve Non-routine Problems?' *Journal of Mathematical Behavior*, **8** (2), 45–50.

Sfard, A., (1989), 'Transition from operational to structural conception: The notion of function revisited' *Proceedings of PME 13*, Paris, **3**, 151–158.

Sierpińska, A., (1985a), 'La notion d'obstacle epistémologique dans l'ensignement des mathematiques', *Proceedings of the 37th CIEAEM Meeting*, Leiden.

Sierpińska, A., (1985b), 'Obstacles epistémologiques relatifs à la notion de limite', *Recherches en Didactique des Mathématiques*, **6** (1), 5–67.

Sierpińska, A., (1987), 'Humanities students and Epistemological Obstacles Related to Limits', *Educational Studies in Mathematics*, **18** (4), 371–87.

Sierpińska, A., (1989), 'How and when attitudes towards mathematics and infinity become constituted into obstacles in students?', *Proceedings of PME 13*, Paris, **3**, 166–173.

Simons, F. H., (1986), 'A course in calculus using a personal computer', *Int. J. Math. Ed. Sci. & Tech.*, **17** (5), 549–552.

Sinclaire, H., (1987), 'Constructivism and the psychology of mathematics', Plenary Paper, *Proceedings of PME 11*, Montréal, **1**, 28–41.

Skemp, R. R., (1971), *The Psychology of Learning Mathematics*, Penguin, London.

Skemp, R. R., (1979), *Intelligence, Learning and Action*, Wiley, London.

Small, D., Hosack, H. & Lane, K. D., (1986), 'Computer Algebra Systems in Undergraduate Instruction', *The College Mathematics Journal*, **17** (5), 423–433.

Southwell, B., (1988), 'Construction and Reconstruction: The Reflective Practice in Mathematics Education', *Proceedings of PME 12*, Veszprem, 584–592.

Stavy, R., Eisen, Y. & Yakobi, D., (1987), 'How students ages 13–15 understand photosynthesis', *International Journal of Science Education*, **9**, 105–115.

Stavy, R., (1991), 'Using analogy to overcome misconceptions about the conservation of matter', *Journal of Research in Science Education*, **28**, 305–313.

Steffe, L., (1988), *Construction of arithmetical meanings and strategies*, Springer-Verlag, New York.

Steiner, M., (1976), *Mathematical Explanation*, Mimeographed Notes, Columbia University.

Stolz, O., (1893), *Grundzüge der Differential und Integral Rechnung*, vol. 1, (ed. Teubner), Leipzig.

Stoutemyer, D. *et al*, (1983), *MuMath*, The Soft Warehouse, Honolulu, Hawaii.

Strauss, S. & Perlmutter, A., (1986), *Teaching a property of the arithmetic average via analogies coded in different symbol systems: A case study in educational–developmental psychology*, unpublished manuscript, Tel-Aviv University, Israel.

Sullivan, K. A., (1976), The teaching of elementary calculus using the non-standard approach, *American Mathematical Monthly*, 370–375.

Sweller, J., (1990), 'On the limited evidence for the effectiveness of teaching general problem solving strategies', *Journal for Research in Mathematics Education*, **21** (5), 411–415.

Tall, D. O., (1977), 'Conflicts and catastrophes in the learning of mathematics', *Mathematical Education for Teaching*, **2** (4), 2–18.

Tall, D. O., (1978), Mathematical thinking and the brain, *Proceedings of PME 2*, Osnabrück, 333–344.

Tall, D. O., (1979), 'Cognitive aspects of proof, with special reference to the irrationality of $\sqrt{2}$,' *Proceedings of PME 3*, Warwick, 203–205.

Tall, D. O., (1980a), 'Looking at graphs through infinitesimal microscopes, windows and telescopes', *Mathematical Gazette*, **64**, 22–49.

Tall, D. O., (1980b), 'Intuitive infinitesimals in the calculus', *Abstracts of short communications, Fourth International Congress on Mathematical Education*, Berkeley, page C5.

Tall, D. O., (1980c), 'The notion of infinite measuring numbers and its relevance in the intuition of infinity', *Educational Studies in Mathematics*, **11**, 271–284.

Tall D. O., (1980d), 'Mathematical intuition, with special reference to limiting processes', *Proceedings of PME 4*, Berkeley, 170–176.

Tall, D. O., (1981), 'Intuitions of infinity', *Mathematics in School*, **10**, 30–33.

Tall, D. O., (1986a), *Building and Testing a Cognitive Approach to the Calculus using Computer Graphics*, Ph.D. Thesis, Mathematics Education Research Centre, University of Warwick.

Tall, D. O., (1986b), *Graphic Calculus I, II, III*, (BBC compatible software), Glentop Press, London.

Tall, D. O., (1986c), 'Lies, Damn Lies and Differential Equations', *Mathematics Teaching*, **114**, 54–57.

Tall, D. O., (1986d), 'Using the computer to represent calculus concepts', *Actes de la 4iéme École d'Été de Didactique des Mathématiques et de l'informatique, Orléans*, Rapport de recherche, IMAG Grenoble, 238–264.

Tall D. O., (1987), 'Constructing the concept image of a tangent', *Proceedings of PME 11*, Montréal, **3**, 69–75.

Tall, D. O., (1989), *Real Functions & Graphs: SMP 16–19*, (for BBC compatible computers), Rivendell Software, prior to publication by Cambridge University Press (1991).

Tall, D. O. (in press), 'The transition to advanced mathematical thinking: functions, limits, infinity and proof', to appear in *The National Council of Teachers of Mathematics Handbook on Research in Mathematics Education*, Reston, Virginia.

Tall, D. O., Blokland, P. & Kok, D., (1990): *A Graphic Approach to the Calculus*, (I.B.M. compatible software), Sunburst, Pleasantville NY.

Tall, D. O. & Thomas, M. O. J., (1989): 'Versatile Learning and the Computer', *Focus on Learning Problems in Mathematics*, **11** (2), 117–125.

Tall, D. O. & Thomas, M. O. J., (1991): 'Encouraging versatile thinking in algebra using the computer', *Educational Studies in Mathematics*, **22** (2), 125–147.

Tall, D. O. & Vinner, S., (1981), 'Concept image and concept definition in mathematics with particular reference to limits and continuity', *Educational Studies in Mathematics*, **12** (2), 151–169.

Tall, D. O. & Winkelmann, B., (1988), ' Hidden algorithms in the drawing of discontinuous functions', *Bulletin of the I.M.A.*, **24**, 111–115.

Thom, R., (1971), 'Modern mathematics: an educational and philosophical error?', *American Scientist*, **59**, 695–699.

Thomas, H. L., (1969), *An analysis of stages in the attainment of a concept of function*, Dissertation Abstracts, 30A, 4163.

Thomas, M. O. J. & Tall, D. O., (1988), 'Longer Term Effects of the Use of the Computer in the Teaching of Algebra', *Proceedings of PME 12*, Veszprem, 601–608.

Thomas, M. O. J., (1988), *A conceptual approach to the early learning of algebra using a computer*, (Doctoral dissertation), University of Warwick, U.K.

Thompson, P. W. & Dreyfus, T., (1988), 'Integers and algebra: Parallels in operations of thought', *Journal for Research in Mathematics Education*, **19** (2), 115–133.

Thompson, P. W., (1985a), 'Experience, problem solving and learning mathematics: considerations in developing mathematics curricula', in E. Silver (Ed.), *Teaching and Learning Mathematical Problem Solving*, Erlbaum, Hillsdale NJ, 189–236.

Thompson, P. W., (1985b), 'Computers in research on mathematical problem solving', In E. Silver (Ed.) *Teaching and Learning Mathematical Problem Solving*, Erlbaum, Hillsdale NJ, 417–436.

Thurston, W. P., (1990), 'Mathematical Education', *Notices of the American Mathematical Society*, **37** (7), 844–850.

Tirosh, D., (1985), *The intuition of infinity and its relevance for mathematics education*, unpublished doctoral dissertation, Tel Aviv University, Israel.

Tymoczko, T., (1986), 'Making room for mathematicians in the philosophy of mathematics', *The Mathematical Intelligencer*, **8** (3), 44–50.

Ulam, S. M., (1976), *Adventures of a mathematician*, Scribner, New York.

Van Dalen, D. & Monna, A. F., (1972), *Sets and Integration, An outline of the development*, Wolters-Noordhof, Groningen.

Van Hiele, P. M., (1959), 'La pensée de l'enfant et la géometrie', *Bulletin de l'Association des Professeurs Mathématiques de L'Enseignement Public*, 198.

Van Lehn, K., (1980), 'Bugs are not enough: empirical studies of bugs, impasses, and repairs in procedural skills', *The Journal of Mathematical Behavior*, **3** (2), 3–71.

Vergnaud, G., (1982), Quelques orientations théoriques et méthodiques des recherches françaises en didactique des mathématiques, *Recherches en Didactique des Mathématiques*, **2** (2).

Vinner, S., (1982), 'Conflicts between definitions and intuitions: the case of the tangent', *Proceedings of PME 6*, Antwerp, 24–28.

Vinner, S., (1983), 'Concept definition, concept image and the notion of function', *International Journal of Mathematical Education in Science and Technology*, **14**, 239–305.

Vinner, S., (1988), 'Visual Considerations in College Calculus – Students and Teachers', *Theory of Mathematics Education Proceedings of the Third International Conference*, Antwerp, 109–116.

Vinner, S. & Dreyfus T., (1989), 'Images and Definitions for the Concept of Function', *Journal for Research in Mathematics Education*, **20** (4), 356–366.

Voss, A., (1899, 1916), 'Differential und Integralrechnung', *Encyclopädie der Mathematischen Wissenschaften mit einschluss ihrer anwendungen*, vol II 1.1, 54–134, Leipzig, (ed. Teubner), French translation by J. Molk: 'Calcul Différentiel', *L'Encyclopédie des Sciences Mathématiques*, vol II–1, 242–297.

Wagner, S., (1981), 'Conservation of equation and function under transformations of variable', *Journal for Research in Mathematics Education*, **12** (2), 107–118.

Wagner, S., Rachlin S. & Jensen, R., (1984), *Algebra Learning Project: Final Report*, Dept. of Math Education, University of Georgia, Athens.

Wheeler, M. M. & Martin, G., (1988), 'Explicit knowledge of infinity', *Proceedings of the 10th Annual Meeting of the North American Chapter of PME*, Northern Illinois University, Dekalb, Illinois, 312–318.

White, A., (1988), *Proceedings of the Second Conference on Humanistic Mathematics; Newsletter No.2*, (Available from author, Dept. of Mathematics, Harvey Mudd College, Claremont, California 91711).

Wilder, R., (1967), 'The role of axiomatics in mathematics', *American Mathematical Monthly*, **74**, 115–27.

Wille, F., (1984), *Humor in der Mathematik*, Vandenhoeck und Ruprecht, Göttingen, 84–87.

INDEX

Mathematics Education Library

Managing Editor: A.J. Bishop, Cambridge, U.K.

1. H. Freudenthal: *Didactical Phenomenology of Mathematical Structures.* 1983
   ISBN 90-277-1535-1; Pb 90-277-2261-7
2. B. Christiansen, A. G. Howson and M. Otte (eds.): *Perspectives on Mathematics Education.* Papers submitted by Members of the Bacomet Group. 1986.
   ISBN 90-277-1929-2; Pb 90-277-2118-1
3. A. Treffers: *Three Dimensions.* A Model of Goal and Theory Description in Mathematics Instruction – The Wiskobas Project. 1987 ISBN 90-277-2165-3
4. S. Mellin-Olsen: *The Politics of Mathematics Education.* 1987
   ISBN 90-277-2350-8
5. E. Fischbein: *Intuition in Science and Mathematics.* An Educational Approach. 1987
   ISBN 90-277-2506-3
6. A.J. Bishop: *Mathematical Enculturation.* A Cultural Perspective on Mathematics Education. 1988
   ISBN 90-277-2646-9; Pb (1991) 0-7923-1270-8
7. E. von Glasersfeld (ed.): *Radical Constructivism in Mathematics Education.* 1991
   ISBN 0-7923-1257-0
8. L. Streefland: *Fractions in Realistic Mathematics Education.* A Paradigm of Developmental Research. 1991
   ISBN 0-7923-1282-1
9. H. Freudenthal: *Revisiting Mathematics Education.* China Lectures. 1991
   ISBN 0-7923-1299-6
10. A.J. Bishop, S. Mellin-Olsen and J. van Dormolen (eds.): *Mathematical Knowledge: Its Growth Through Teaching.* 1991
    ISBN 0-7923-1344-5
11. D. Tall (ed.): *Advanced Mathematical Thinking.* 1991 ISBN 0-7923-1456-5
12. R. Kapadia and M. Borovcnik (eds.): *Chance Encounters: Probability in Education.* 1991
    ISBN 0-7923-1474-3

KLUWER ACADEMIC PUBLISHERS – DORDRECHT / BOSTON / LONDON